Lower vertebrates from the Palaeozoic

First International Palaeontological Congress
(IPC 2002), Sydney, Australia, July 2002

Proceedings of Symposium 6
[Palaeozoic Vertebrates]

Edited by Gavin C. Young

Contents

Preface

The bold idea of instigating a series of regular international congresses, under the auspices of the International Palaeontological Association (IPA), was realised with the First International Palaeontological Congress (IPC 2002), held in Sydney, Australia, in July 2002. These are to be held at 4 yearly intervals, and the Second International Palaeontological Congress is scheduled for Beijing, China, in 2006.

IPC 2002 was hosted by the Macquarie University Centre for Ecostratigraphy and Palaeobiology (MUCEP), and the Australian Museum, Sydney. It was sponsored locally by the Association of Australasian Palaeontologists (AAP). Formal sessions took place at Macquarie University. The First International Palaeontological Congress brought together more than 400 palaeontologists from 35 nations. In total, 199 oral papers and 172 posters were presented in some 24 symposia (held in four parallel sessions), covering all aspects of palaeontology and palaeobiology, and there was a consensus that this was one of the most stimulating scientific meetings of recent years. IPC 2002 achieved its objectives by demonstrating that palaeontology in all its guises remains a vigorous scientific pursuit, replete with new ideas, and expanding its linkages across an increasingly broad spectrum of the sciences. Some 200 participants also experienced the marvellous and unique fossil sequences of Australia and New Zealand in a range of pre-, mid- and post-congress excursions, from the Ediacaran of the Flinders Ranges in South Australia, to the Cainozoic vertebrate faunas of the Riversleigh World Heritage area, and from the Devonian reef complexes of the Canning Basin, Western Australia, to contemporary reef dynamics at Heron Island on the Great Barrier Reef.

In the field of early vertebrates, Australia has some occurrences of international significance, including the world's oldest known fish (*Arandaspis* from the Ordovician of central Australia), and the unequalled preservation of acid-prepared Upper Devonian fishes from Gogo in Western Australia. It was appropriate, therefore, to organise a symposium on "Palaeozoic Vertebrates" for IPC 2002. Participants from seven countries contributed 17 oral presentations and 12 posters to this symposium. Systematics, biostratigraphy, biogeography, taphonomy and other aspects were covered for a range of early vertebrate groups, including jawless agnathans, the early jawed fishes (acanthodians, placoderms, sharks, lobe-finned fishes), and Palaeozoic tetrapods. The collection of papers published here results from the contributions to that symposium.

Extensive fossil fish assemblages occur in Devonian marine and non-marine strata across much of central and eastern Australia. They represent the first vertebrate fauna to occupy the Australian continent, but are still largely undescribed, some 150 years after such remains were first documented in the scientific literature of Europe. During the Middle Palaeozoic, Australia and Antarctica formed a major area of endemism, the "East Gondwana Province", and its biogeography has recently been summarised as part of a major overview of fossil floras and faunas from the Phanerozoic of Australasia (Wright *et al.* 2000: Palaeobiogeography of Australasian faunas and floras, *Association of Australasian Palaeontologists, Memoir 23*, 515 pp.).

This proceedings volume includes some major contributions to the systematic documentation of Devonian vertebrate faunas from both Australia and Antarctica. Such research underpins the study of distribution patterns in time and space, by which palaeontological data can be applied to questions of biogeography and palaeogeography. Thus, these papers are a contribution to the recently approved IGCP Project 491 on "Middle Palaeozoic vertebrate biogeography, palaeogeography and climate". Inevitably, considerations of phylogeny, biogeography, biostratigraphy and taphonomy require a global perspective, and other contributions cover these aspects for Northern Hemisphere regions, ranging from the Siluro-Devonian of Spitsbergen and Siberia, to the Permian of Germany. These published papers give an indication of the broad coverage of early vertebrate research presented at IPC 2002.

IPC 2002 would not have happened without the drive and commitment of its convenor, the then IPA President, Professor John Talent (Earth and Planetary Sciences, Macquarie University). He was supported by a wonderful team of MUCEP staff, research associates, students, and friends at Macquarie University, who organised the scientific programme, and associated programmes for science teachers and accompanying persons, and managed registrations, finances, accommodation, displays, transport, publicity, and all the other time-consuming demands of a large scientific conference.

As convenor of Symposium 6 (Palaeozoic Vertebrates) I would like to thank all contributors for the quality of their presentations, and as organiser of the post-congress vertebrate excursion I must acknowledge the assistance of Alex Ritchie, Carole Burrow and Lynne Bean for the 7 days we spent showcasing the significant Devonian fish/tetrapod localities of southeastern Australia to 17 of our overseas colleagues. As editor of this volume, I thank the contributing authors for their efforts, and the following experts in early vertebrate research for their penetrating and timely reviews of submitted manuscripts: K. S. W. Campbell, E. Daeschler, D. Elliott, P.-Y. Gagnier, G. Hanke, Z. Johanson, H. Johnson, A. Kemp, J. A. Long, E. Mark-Kurik, M. Richter, H.-P. Schultze, S. Turner, J. Valiukevicius, J. Vergoossen, and A. Warren.

Finally, on behalf of all contributors, I express my gratitude to Professor David L. Bruton and the Lethaia Foundation for supporting the publication of these proceedings.

Gavin C. Young
Convenor of Symposium 6 (Palaeozoic Vertebrates),
IPC 2002
Guest Editor

491
IUGS
UNESCO

Biostratigraphy of Pteraspidiformes (Agnatha, Heterostraci) from the Wood Bay Formation, Lower Devonian, Spitsbergen

VINCENT N. PERNEGRE

Pernegre, V.N. **2004 06 01**: Biostratigraphy of Pteraspidiformes (Agnatha, Heterostraci) from the Wood Bay Formation, Lower Devonian, Spitsbergen. *Fossils and Strata*, No. 50, pp. 1–7. France. ISSN 0300-9491.

A new study of the pteraspidiform fauna of the Wood Bay Formation, which began in 1999 at the Muséum National d'Histoire Naturelle de Paris, has resulted in an improvement in our current knowledge of this fauna by documenting an increased diversity of forms representative of the three main families of Pteraspidiformes: Protopteraspididae, Pteraspididae and Protaspididae. Previous published works and original data are used to revise preceding biostratigraphic work on the Wood Bay Formation. Spitsbergen is compared with other circum-Arctic regions, mainly Severnaya Zemlya and Novaya Zemlya, and possible biostratigraphic correlations are established.

Key words: Doryaspis; Gigantaspis; Spitsbergaspis; stratigraphic fossils; biostratigraphy; Lower Devonian.

Vincent N. Pernegre [pernegre@hotmail.com], USM 0203, Département Histoire de la Terre, Muséum National d'Histoire Naturelle, 8 rue Buffon, 75005 Paris, France

Introduction

Pteraspidiform heterostracans represent a substantial part of the palaeontological material in the Wood Bay Formation of Spitsbergen. They were first mentioned by Lankester (1884), but were actually only studied in the second part of the 20th century (Heintz 1960, 1962, 1967; Blieck & Goujet 1983).

The Wood Bay Formation in Spitsbergen (Fig. 1A, B) represents the upper part of the Lower Devonian. Its age is commonly estimated from vertebrates and miospores to be Pragian–Emsian (Blieck & Cloutier 2000). It is composed of four faunal divisions (Goujet 1984), which are, in order from oldest to youngest: Sigurdfjellet, Kapp Kjeldsen, Keltiefjellet (often presented as equivalent to the "Lykta" fauna), and Stjørdalen. These faunal divisions were initially proposed and defined by Friend *et al.* (1966) and completed by Goujet (1984) who added the lower one (Sigurdfjellet).

From an extensive study of material collected during the CNRS-MNHN expedition of 1969, the genus *Doryaspis* has been revised (Pernegre 2002), a new genus has been described as *Spitsbergaspis* Pernegre, 2003, and the genus *Gigantaspis* will be revised soon. Pteraspidiform biostratigraphy of the Wood Bay Formation has been revisited and refined in the light of this new information.

Biostratigraphy

History

Føyn & Heintz (1943) divided the Wood Bay Formation into three divisions which can be paleontologically and geologically distinguished. "Each of these three divisions is characterised by a group of guide-fossils. With our present knowledge of the stratification and of the fossils it is therefore impossible to draw distinct limits between the various divisions … In appearance all three divisions have certain (geological) characteristics making it possible to distinguish them more or less accurately at a distance" (p. 13).

Friend (1961, p. 81) commented that "many of the vertebrates are only represented in collections by isolated fragments and have not yet been adequately described". He proposed a new lithological description of the divisions for field geologists, based on "simple field criteria such as grain-size or colour of the sediment".

Friend *et al.* (1966) pointed out the "unsatisfactory nature" of these divisions in lithostratigraphic terms, and proposed replacing them with the concept of "faunal divisions", using new observations on the vertebrate fauna of the Wood Bay Formation (Heintz 1960, 1962).

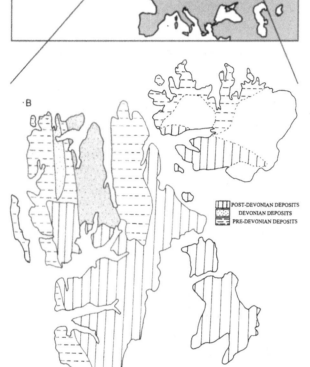

Fig. 1.   A: Locality map of the Svalbard Archipelago. B: Outcrop of the Wood Bay Formation (Lower Devonian, in grey) in Spitsbergen.

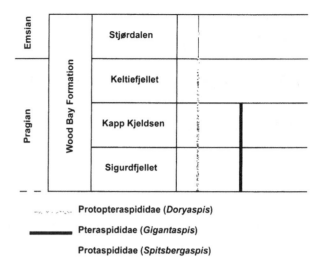

Fig. 2.   Stratigraphic table of the Wood Bay Formation (Lower Devonian, Spitsbergen), showing the ranges of presently known pteraspidiform genera.

recorded a large diversity of representatives of *Doryaspis*, and some undescribed new forms of other genera (*Gigantaspis* and *Zascinaspis*).

Current research

Recent investigations have complemented the previous works by providing new and more detailed stratigraphical distributions of forms, and identifying new pteraspidiform taxa for formal systematic study. Currently, three genera of Pteraspidiformes have been identified, and their distribution documented through the Wood Bay Formation (Fig. 2). The typical representative of this formation is *Doryaspis* Lankester, 1886 (Protopteraspididae), which occurs in all faunal divisions (Fig. 2) with at least five identified species (one new, see below). This taxon characterises the Wood Bay Formation as it does not occur elsewhere. The other well-represented genus is *Gigantaspis* Heintz, 1962 (Pteraspididae), which is restricted to the first two faunal divisions (Sigurdfjellet and Kapp Kjeldsen; Fig. 2), and has four identified species (one new, see below). Finally, the genus *Spitsbergaspis* Pernegre, 2003 (Protaspididae) occurs only in the basal faunal division of the Wood Bay Formation (Sigurdfjellet; Fig. 2) and is represented by a single species. A brief description of the Spitsbergen genera follows.

Doryaspis (Fig. 3) is characterised by a flattened dorsal shield and a swollen ventral one. It possesses a ventral elongated pseudo-rostrum (a modified median oral plate; Pernegre 2002) and bears laterally a pair of well-developed cornual plates. In this genus, the rostrum is represented by a small truncated and arched plate which delimits the dorsal part of the mouth (Heintz 1967).

The representatives of *Gigantaspis* (Fig. 4) are dorsoventrally flattened fishes with long branchial plates characterised by a very narrow dorsal lamella. Most of the

These faunal divisions were mainly based on the pteraspidiform heterostracans. It was accepted that the genus *Gigantaspis*, "with a small species of *Doryaspis*", characterised the Kapp Kjeldsen division (Friend *et al.* 1966, p. 61) and the Keltiefjellet division was characterised by *Doryaspis nathorsti*, and the absence of *Gigantaspis*.

Goujet (1984) distinguished the basal layers of the Kapp Kjeldsen division in an independent faunal division, Sigurdfjellet. The distinctive fauna at this level was first pointed out by Føyn & Heintz (1943), and was enhanced by a study of the arthrodires (Goujet 1973).

Palaeontological diversity was considered to be low until Blieck *et al.* (1987) presented a biostratigraphic table of the Wood Bay Formation, which indicated a great diversity of forms in all the now recognised four faunal divisions, but with many undescribed taxa. These authors

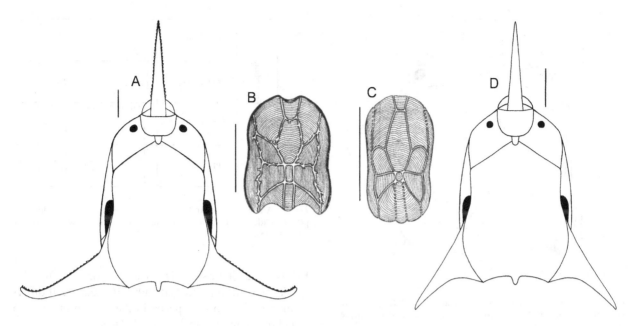

Fig. 3. Reconstructions of described species of the genus *Doryaspis* (Wood Bay Formation, Lower Devonian, Spitsbergen). A: Dorsal shield of the type species *Doryaspis nathorsti*, after Pernegre (2002). B: Dorsal disc of *Doryaspis lyktensis*, after Heintz (1960). C: Dorsal disc of *Doryaspis minor*, after Heintz (1960). D: New reconstruction of the dorsal shield of *Doryaspis arctica*, after Pernegre (2002). Scale bars = 1 cm.

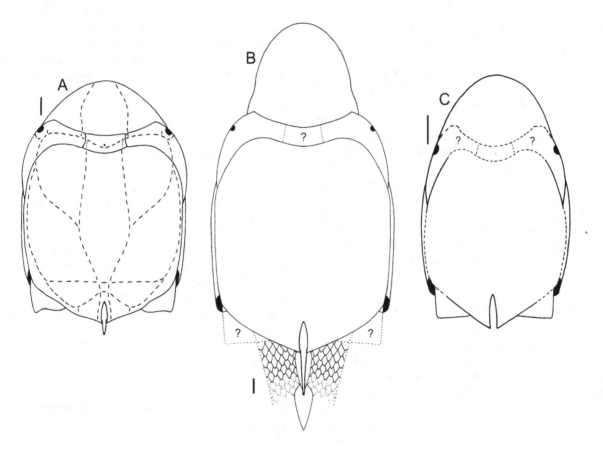

Fig. 4. Reconstructions of described species of the genus *Gigantaspis* (Wood Bay Formation, Lower Devonian, Spitsbergen). A: Dorsal shield of *Gigantaspis laticephala*, after Blieck & Goujet (1983). B: New reconstruction of the dorsal shield of *Gigantaspis isachseni*. C: Dorsal shield of *Gigantaspis bocki*, after Heintz (1962) and Blieck (1984). Scale bars = 1 cm.

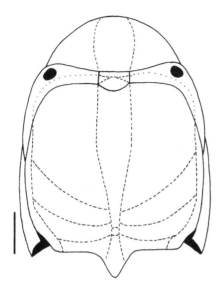

Fig. 5. Reconstruction of the dorsal shield of *Spitsbergaspis prima* (Wood Bay Formation, Lower Devonian, Spitsbergen), after Pernegre (2003). Scale bar = 1 cm.

Fig. 6. A new detailed stratigraphy of species of the Pteraspidiformes presently identified in the Wood Bay Formation (Lower Devonian, Spitsbergen).

species possess a pair of cornual plates developed at the posterior tip of the dorsal shield.

Spitsbergaspis (Fig. 5) is characterised by its dermal ornamentation, which varies along both dorsal and ventral discs, with smooth dentine ridges in the central area and tubercles around the lateral margins (Pernegre 2003). This form is also dorsoventrally flattened, and shows no cornual plates and no dorsal spine.

From an extensive bibliographical review, and study of collections of the known pteraspidiform faunal localities in the Wood Bay Formation (Nathorst expeditions, ENS 1939, MNHN-CNRS 1969; Lankester 1884; White 1935; Føyn & Heintz 1943; Heintz 1960, 1962, 1967; Goujet 1984; Blieck 1984; Blieck *et al.* 1987), each species of the three preceding genera can be stratigraphically well delimited. Their different occurrences and associations allow us to characterise and differentiate each faunal division of the Wood Bay Formation precisely.

Occurrence of the five *Doryaspis* species

The first species of *Doryaspis* appears in the basal faunal division of the Wood Bay Formation (Sigurdfjellet; Fig. 6). *Doryaspis arctica* (Fig. 3D) occurs in abundance at all known localities of the Sigurdfjellet division. It is mainly represented by isolated plates, but some articulated specimens have been found (Pernegre 2002). This species also occurs in some localities at the base of the Kapp Kjeldsen division (Fig. 6) in association with its characteristic fossil *Gigantaspis isachseni* (see below).

The second species is found in only one locality at Groenhorgdalen, in association with *Gigantaspis laticephala* (see below). According to Føyn & Heintz (1943), these sediments belong to the upper part of the Kapp

Kjeldsen faunal division in the Dicksonfjord (Fig. 6). This species remains undescribed (submitted paper); it differs from the other known species in the pattern of its main shield plates and its small cornual plates.

The third species is the type species *Doryaspis nathorsti* (Fig. 3A). It occurs abundantly in the localities of the Keltiefjellet faunal division (Fig. 6), sometimes almost completely articulated (Pernegre 2002). Heintz (1967) noted that some specimens could also have come from the upper part of the Kapp Kjeldsen division. During our present study, only one specimen in the Palaeontological Museum, University of Oslo collection confirms this hypothesis; a right cornual plate of *Doryaspis nathorsti* associated with a rostrum of *Gigantaspis isachseni* (specimen PMO A41030).

Within the same Keltiefjellet division (Fig. 6), we can find another associated species: *Doryaspis lyktensis* (Fig. 3B). This very small species (Heintz 1960), with dorsal discs 1–2 cm long, is not abundant, and is always disarticulated. It is known only from some dorsal and ventral discs.

The last species of the genus is the poorly preserved *Doryaspis minor* (Fig. 3C), the smallest species described by Heintz (1960), who recorded it from the highest part of the Wood Bay Formation – the Stjørdalen faunal division (Fig. 6). Presently only one specimen has been identified in the Palaeontological Museum, University of Oslo (PMO A27842), and no additional material has yet been found in the collected localities from the 1969 expedition.

Occurrence of the four *Gigantaspis* species

Stratigraphically, the first occurring species of this genus is a new one (description in progress by the author), which appears in the Sigurdfjellet faunal division (Fig. 6). It is the smallest species of the genus, with a dorsal disc about 6 cm long, and can be compared in its general form with *Gigantaspis laticephala* (see below), in its small rostrum and wide pineal plate, with cornual plates at the posterior tip of the shield. This new species, mentioned by Blieck *et al.* (1987), is well represented (but no articulated specimens were found) in all localities of the Sigurdfjellet division, often in association with *Doryaspis arctica*.

In the overlying Kapp Kjeldsen division (Fig. 6), it is replaced by the largest and type species *Gigantaspis isachseni* (Fig. 4B). This form is characterised by its triangular rostral plate which is as long as wide, and the dimensions of its dorsal discs (Heintz 1962), often more than 20 cm long and 15 cm wide. *Gigantaspis isachseni* is abundantly represented, sometimes in association with a less abundant species, *Gigantaspis bocki* (Figs. 4C, 6). This species, smaller than *Gigantaspis isachseni*, is poorly known, but can be considered as the most elongated species of the genus.

In the upper part of the Kapp Kjeldsen faunal division is found the last species of the genus, *Gigantaspis laticephala* (Fig. 4A), known only from a single locality, Groenhorgdalen, Dicksonfjord. It was originally described and assigned to the genus *Zascinaspis* by Blieck & Goujet (1983), but in fact does not differ from the genus *Gigantaspis*. This is the flattest species of the genus, and is always associated with the new species of *Doryaspis* mentioned above (Fig. 6).

Occurrence of *Spitsbergaspis*

Only one species is known for this genus: *Spitsbergaspis prima* Pernegre, 2003 (Fig. 5). This species occurs in many localities of the Sigurdfjellet faunal division, and is often associated with the small new species of *Gigantaspis*, and sometimes with *Doryaspis arctica* (Fig. 6).

Characteristic associations and biostratigraphic indicators

Given the preceding data on the stratigraphic occurrence of the Pteraspidiformes in the Wood Bay Formation, we can check and document each faunal division, using stratigraphic fossils or more often some associations.

The Sigurdfjellet sediments are characterised by the presence of *Spitsbergaspis prima* associated with the new small *Gigantaspis* species and *Doryaspis arctica*. The occurrence of *Doryaspis arctica* on its own is not sufficient to identify the Sigurdfjellet division, whereas the presence of either *Spitsbergaspis prima* and/or the small species of *Gigantaspis* is sufficient.

The following Kapp Kjeldsen faunal division can be characterised at different levels. The occurrence of *Gigantaspis isachseni* allows us to delimit the entire

division (stratigraphic fossil), sometimes in association with *Gigantaspis bocki*. The association of *Gigantaspis isachseni* with *Doryaspis arctica* characterises the first half of the Kapp Kjeldsen division. In the south part of the graben, the association of *Gigantaspis laticephala* and the new species of *Doryaspis* may characterise, for the Dicksonfjord area, the upper part of the Kapp Kjeldsen division.

In the overlying Keltiefjellet faunal division, *Doryaspis nathorsti* is the characteristic index fossil. This species could appear as early as the top of the Kapp Kjeldsen division, but presently only one specimen supports this hypothesis (see above). *Doryaspis nathorsti* is represented in all localities of the Keltiefjellet division, often in association with the small species *Doryaspis lyktensis*. The genus *Doryaspis* is the only pteraspidiform occurring in this faunal division and in the following Stjørdalen.

The Stjørdalen faunal division is the highest division of the Wood Bay Formation. It possesses a poor pteraspidiform fauna (Blieck *et al.* 1987), currently only known in one locality, and represented by *Doryaspis minor*. This form certainly characterises a part of the division, but its occurrence throughout the division is presently doubtful due to the lack of sufficient material in collections, and the scarcity of this form in localities.

Biostratigraphic comparisons

For a long time, the Pteraspidiformes of the Wood Bay Formation were considered to be endemic to Spitsbergen. However, Mark-Kurik & Novitskaya (1977, p. 149) noted and illustrated some pteraspidiform remains in the Lower Devonian of Novaya Zemlya, which they considered "quite similar to *Gigantaspis* ... [and to] the aberrant pteraspids" (currently known as *Doryaspis*). These conclusions were based on the similar size and ornamentation of the Novaya Zemlya forms with those already known in Spitsbergen, and the authors proposed an equivalence in age with the Kapp Kjeldsen division of Spitsbergen.

Karatajute-Talimaa (1983, p. 23) quoted the occurrence of both *Gigantaspis* and ?*Doryaspis* as rare vertebrate remains from the Lower Devonian strata of Severnaya Zemlya archipelago (Spokojnaya Formation). Recently Blieck *et al.* (2002) confirmed the presence of *Gigantaspis* and a member of the Protopteraspididae with a *Doryaspis*-like ornamentation in the Spokojnaya Formation from October Revolution Island. They concluded that the Spokojnaya Formation is an equivalent of the base of the Wood Bay Formation, without any precision of faunal division.

The new data on the stratigraphic subdivision of pteraspidiform occurrences in the Wood Bay Formation given above indicate that the Spokojnaya Formation

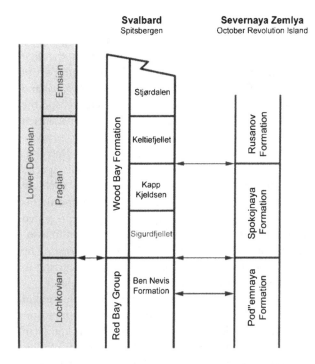

Fig. 7. Proposed biostratigraphic correlations between the Lower Devonian of Spitsbergen and Severnaya Zemlya (Novaya Zemlya not taken into account due to the lack of precise subdivisions of its Lower Devonian sequence).

from Severnaya Zemlya is equivalent to the associated Sigurdfjellet and Kapp Kjeldsen divisions (Fig. 7). All of them contain both *Gigantaspis* and *Doryaspis*, or close representatives. The unnamed part of the Lower Devonian from Novaya Zemlya (Karatajute-Talimaa 1983), containing representatives of these genera, can be correlated with the same faunal divisions of Spitsbergen, and not only with the Kapp Kjeldsen division (Fig. 7), due to the introduction of the Sigurdfjellet faunal division by Goujet (1984) at the base of the Wood Bay Formation.

Conclusion

The new information presented here increases our knowledge of the biodiversity of Pteraspidiformes in the Lower Devonian of Spitsbergen, with the identification of two new species: *Doryaspis* sp. nov. (Kapp Kjeldsen division) and *Gigantaspis* sp. nov. (Sigurdfjellet division). The study and revision of pteraspidiform taxa in all the known field localities allows the precise documentation of a new scheme of Wood Bay Formation biostratigraphy.

The material demonstrates a great diversity of forms belonging to three genera, each representing one of the main pteraspidiform families: Protopteraspididae, Pteraspididae and Protaspididae. This material allows biostratigraphical correlations with other circum-Arctic

regions such as Severnaya Zemlya and Novaya Zemlya. These correlations will be a starting point for more enhanced circum-Arctic faunal comparisons, to be supported when the fauna from the Siberian archipelagos is formally described.

Acknowledgements

I am grateful to Professor D. Goujet for facilitating the study of material in the collections of the Muséum National d'Histoire Naturelle, Paris, and for arranging grants to cover my participation at the First International Palaeontological Congress in Sydney. I thank G. Young and A. Ritchie for their friendly reception in Australia, during the congress and the following field trip, and D. Elliott for his critical revision of my original manuscript.

References

Blieck, A. 1984: Les Hétérostracés Ptéraspidiformes. Systématique, phylogénie, biogéographie. *Cahiers de Paléontologie (section Vertébrés)*. CNRS. 199 pp.

Blieck, A. & Cloutier, R. 2000: Biostratigraphical correlations of Early Devonian vertebrate assemblage of the Old Red Sandstone Continent. *In* Blieck, A. & Turner, S. (eds): *Palaeozoic Vertebrate Biochronology and Global Marine/Non-marine Correlation, Final Report of the IGCP 328. Courier Forschungsinstitut Senckenberg 223.* Frankfurt.

Blieck, A. & Goujet, D. 1983: *Zascinaspis laticephala* nov. sp. (Agnatha, Heterostraci) du Dévonien inférieur du Spitsberg. *Annales de Paléontologie 69*, 43–56.

Blieck, A., Goujet, D. & Janvier, P. 1987: The vertebrate stratigraphy of the Lower Devonian (Red Bay Group and Wood Bay Formation) of Spitsbergen. *Modern Geology 11*, 197–217.

Blieck, A., Karatajute-Talimaa, V.N. & Mark-Kurik, E. 2002: Upper Silurian and Devonian heterostracan pteraspidomorphs (Vertebrata) from Severnaya Zemlya (Russia): a preliminary report with biogeographical and biostratigraphical implications. *Geodiversitas 24(4)*, 805–820.

Føyn, S. & Heintz, A. 1943: The English–Norwegian–Swedish expedition 1939. Geological results. The Downtonian and Devonian vertebrates of Spitsbergen VIII. *Norges Svalbard- og Ishavs-Undersøkesler, Skrifter 85*, 1–51.

Friend, P.F. 1961: The Devonian stratigraphy of north and central Vestspitsbergen. *Proceedings of the Yorkshire Geological Society 33*, 77–118.

Friend, P.F., Heintz, N. & Moody-Stuart, M. 1966: New unit terms for the Devonian of Spitsbergen and new stratigraphical scheme for the Wood Bay Formation. *Norsk Polarinstitutt, Årbok 1965*, 59–64.

Goujet, D. 1973: *Sigaspis*, un nouvel Arthrodire du Dévonien inférieur du Spitsberg. *Palaeontographica A 143*, 73–88.

Goujet, D. 1984: Les poissons placodermes du Spitsberg. Arthrodires Dolichothoraci de la Formation de Wood Bay (Dévonien inférieur). *Cahiers de Paléontologie (section Vertébrés)*. CNRS. 254 pp.

Heintz, N. 1960: Two new species of the genus *Pteraspis* from the Wood Bay Series in Spitsbergen. The Downtonian and Devonian vertebrates of Spitsbergen X. *Norsk Polarinstitutt, Skrifter 117*, 3–13.

Heintz, N. 1962: *Gigantaspis* – a new genus of family Pteraspididae from Spitsbergen. The Downtonian and Devonian vertebrates of Spitsbergen XI. *Norsk Polarinstitutt, Årbok 1960*, 22–27.

Heintz, N. 1967: The Pteraspid *Lyktaspis* n.g. from the Devonian of Vestspitsbergen. *In* Ørvig, T. (ed.): *Current Problems of Lower Vertebrate Phylogeny. Nobel Symposium IV*, 73–80.

Karatajute-Talimaa, V.N. 1983: The Lower Devonian heterostracans from Severnaya Zemlya and their importance for correlations. *In* Novitskaya, L.I. (ed.): *Extant Problems of Paleoichthyology*, 22–28. Nauka, Moscow.

Lankester, E.R. 1884: Report on fragments of fossil fishes from Palaeozoic strata of Spitsbergen. *Kungliga Ventenskaps-Akademiens Handlingar 20*, 1–6.

Mark-Kurik, E. & Novitskaya, L. 1977: The Early Devonian fish-fauna on Novaja Zemlya. *Eesti NSV Teaduste Akadeemia Toimetised, Geoloogia 26*, 143–149.

Pernegre, V.N. 2002: The genus *Doryaspis* White (Heterostraci) from the Lower Devonian of Vestspitsbergen, Svalbard. *Journal of Vertebrate Paleontology 22*, 735–746.

Pernegre, V.N. 2003: Un nouveau genre de Pteraspidiforme (Vertebrata, Heterostraci) de la Formation de Wood Bay (Dévonien inférieur, Spitsberg). *Geodiversitas 25*, 261–272.

White, E.I. 1935: The Ostracoderm *Pteraspis* Kner and the relationships of the agnathous vertebrates. *Philosophical Transactions of the Royal Society of London 225*, 381–457.

Acanthodian fishes with dentigerous jaw bones: the Ischnacanthiformes and *Acanthodopsis*

CAROLE J. BURROW

Burrow, C.J. **2004 06 01**: Acanthodian fishes with dentigerous jaw bones: the Ischnacanthiformes and *Acanthodopsis*. *Fossils and Strata*, No. 50, pp. 8–22. Australia. ISSN 0300-9491.

Within the early gnathostome group Acanthodii, several different types of dentition are exhibited. Of the Siluro-Devonian acanthodians, some of the Climatiidae and the Brochoadmonidae have rows of small tooth whorls lining their jaws, while the Ischnacanthidae and Poracanthodidae have dermally derived dentigerous jaw bones, with some taxa also having a large symphysial tooth whorl and smaller tooth whorls and teeth within the mouth cavity. A change in biting mode from a vertically occluding, cog-like action in Silurian–Early Devonian fish to a shearing action in Middle–Late Devonian fish is indicated by a change in the layout, shape and wear of the teeth on the jaw bones. *Acanthodopsis*, a Carboniferous genus of the Acanthodidae, also has dentigerous jaw bones. A revised generic diagnosis of this form is presented, and isolated jaws of a new Australian species, *Acanthodopsis russelli* sp. nov., give further evidence that these dentigerous elements are not homologous with those of ischnacanthiforms. Some new examples of *Ischnacanthus gracilis* jaws are illustrated, and compared with those of *Acanthodopsis*, in which the bone is a perichondral ossification that covers the entire jaw cartilage, without a separate, toothed dermal bone forming the biting surface.

Key words: Acanthodidae; Acanthodii; *Acanthodopsis*; Devonian; Carboniferous; Ischnacanthiformes; dentigerous jaw bones.

Carole J. Burrow [C.Burrow@uq.edu.au], Department of Zoology and Entomology, University of Queensland, QLD 4072, Australia

Introduction

The Acanthodii are one of the first gnathostome groups to appear in the fossil record, with the oldest known remains being from the Late Ordovician/Early Silurian (Karatajute-Talimaa 1998; Karatajute-Talimaa & Smith 2002). Three orders are "classically" recognised within the Class Acanthodii: the Climatiiformes, Acanthodiformes and Ischnacanthiformes (Berg 1940; he also erected the Mesacanthiformes, Gyracanthiformes, Diplacanthiformes, and Cheiracanthiformes, which have since been reduced to family level). Of these, the Climatiiformes are now considered paraphyletic (Janvier 1996; Hanke & Wilson 2002), while the Acanthodiformes and the Ischnacanthiformes are deemed monophyletic. The loss of one dorsal fin and spine is the main synapomorphy of the Acanthodiformes.

The Ischnacanthiformes are characterised by dentigerous jaw bones of dermal origin, and within the group two families are currently recognised: the Ischnacanthidae,

with the type species *Ischnacanthus gracilis* (Egerton, 1861) based on articulated fish from the Early Devonian of Scotland (Fig. 1A), and the Poracanthodidae, for which the type is *Poracanthodes punctatus* (Brotzen, 1934), a taxon erected for isolated scales from Late Silurian Beyrichienkalk erratic boulders of northern Germany (Fig. 1B, C). Most ischnacanthiform taxa were erected for isolated jaw bones, with very few taxa based on articulated fish.

This paper presents a review of acanthodians with dentigerous jaw bones, including the Carboniferous acanthodid genus *Acanthodopsis*, for which a revised generic diagnosis is presented. New examples of jaw bones of the Scottish form *Ischnacanthus gracilis* are illustrated, and isolated jaws from the Carboniferous of Australia are described as a new species of *Acanthodopsis*. The evidence of this new species supports the view that the dentigerous elements of *Acanthodopsis* are not homologous with those of ischnacanthiforms.

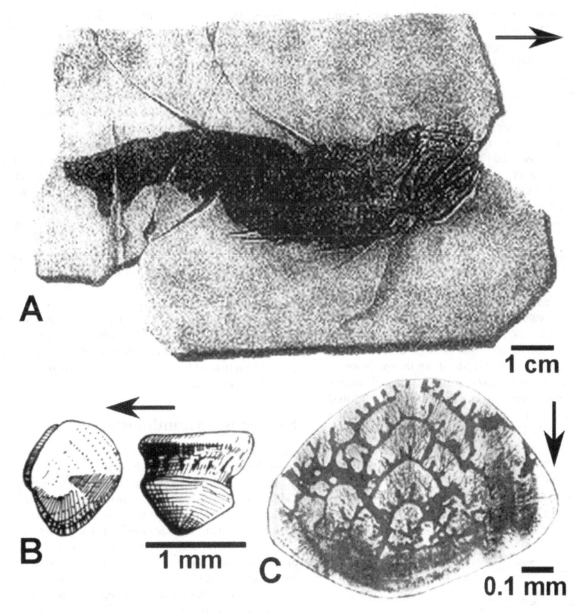

Fig. 1.   A: Holotype specimen of *Ischnacanthus gracilis* NHM P6987 (Egerton 1861, pl. 9, fig. 1), Lower Devonian (Lochkovian) Lower Old Red Sandstone, Farnell, Scotland. B: Holotype scale of *Poracanthodes punctatus* (Brotzen 1934a, pl. 3, fig. 1a, b), Lower Devonian erratic boulder Bey. 36, north Germany. C: Horizontal section through crown of *Poracanthodes punctatus* scale (Brotzen 1934b, pl. 7, fig. 1), possibly from the same sample; type material presumed lost. Arrows are anteriad.

Abbreviations for fossil localities and palaeontological collections housed in various institutions mentioned in the text are: ANU, Australian National University, Canberra; E, Buffalo Museum of Science; NEWHM G, Hancock Museum; MOTH, "Man on the Hill" locality, Northwest Territories, Canada; NHM P, Natural History Museum, London; NMS, National Museum of Scotland; NYSM P, New York State Museum; QM, Queensland Museum; QML, Queensland Museum locality number; QM CB, Carole Burrow collection in the Queensland Museum.

Review of acanthodians with dentigerous jaw bones

As well as the type species *Ischnacanthus gracilis*, other ischnacanthiform taxa based on articulated fish include *Onchus graptolitarum* Fritsch, 1907 from the Silurian of Bohemia, *Zemlyacanthus* (*Poracanthodes*) *menneri* (Valiukevicius, 1992), *Acritolepis ushakovi* Valiukevicius, 2003, *Acritolepis urvantsevi* Valiukevicius, 2003 and *Acanthospina irregulare* Valiukevicius, 2003 from the Early Devonian of Severnaya Zemlya, Arctic Russia, and

Atopacanthus sp. (Jessen 1973) from the early Upper Devonian of Bergisch Gladbach, Germany. Bernacsek & Dineley (1977) assigned specimens from the Lower Devonian of the MOTH locality to *Ischnacanthus gracilis*, but it is now known that there are several different ischnacanthiform taxa in that assemblage, and these are currently under study (Gagnier & Wilson 1995; Hermus & Wilson 2001). Other new forms of articulated ischnacanthiforms are also known from Severnaya Zemlya (Valiukevicius 1997, 2003). These new taxa all have dentigerous jaw bones typical of ischnacanthiforms. Despite this feature, Valiukevicius (2003) assigned *Acritolepis* to the Climatiiformes and *Acanthospina* to Acanthodii *incertae sedis*, based on the morphology and histology of their scales.

Uraniacanthus spinosus Miles, 1973, erected for articulated fish from the Early Devonian siltstones of Wayne Herbert quarry, Herefordshire, was originally described as an ischnacanthiform. Of the type material, the holotype NHM P16609 lacks a head; another specimen NHM P16612-3 (Miles 1973, pl. 13, figs. 1, 2) has the head squashed dorsoventrally, with no sign of dentigerous jaw bones (elements labelled "dg. B" on the specimen are identical to the post-orbital scales of *Gladiobranchus*; G. Hanke 2003, pers. comm.). Specimen NHM P53032 (Miles 1973, pl. 12, fig. 1) was preserved on part of a broken block (NHM P53032-5), and described as comprising "a body wanting the tail and much of the head, but with associated jaws" (Miles 1973, p. 147). These jaws, although associated, are probably assignable to *Ischnacanthus* sp. rather than being from *Uraniacanthus*

spinosus. Hanke *et al.* (2001a, p. 751) mentioned that Sam Davis (UK) is currently studying *Uraniacanthus* "to determine whether *Uraniacanthus* is related to *Gladiobranchus* and the diplacanthids, or to the ischnacanthids". Burrow (1996) tentatively assigned partial articulated fish from the late Givetian/early Frasnian Bunga Beds, New South Wales, to Ischnacanthidae?, but all fish lack heads and so their dentition is unknown. One other genus, *Marsdenius* Wellburn, 1902, was erected for articulated fish (since lost, and never figured) with dentigerous jaw bones, two dorsal fins, and ornamented flank scales from the Lower Carboniferous Pendleside limestone, Yorkshire. Wellburn (1902) compared the *Marsdenius* jaw bones with those of *Acanthodopsis*, for which the type species is *Acanthodopsis wardi* Hancock & Atthey, 1868 from the Upper Carboniferous Coal Measures of Britain. The type specimen of *A. wardi* is a head and pectoral region of a fish, but most specimens assigned to the taxon are isolated upper and lower jaw bones. *Acanthodopsis* has been assigned either to the Ischnacanthiformes because the jaw bones are dentigerous, or to the Acanthodiformes based on the acanthodiform-type double mandibular joint. The relationships of this genus are discussed later in the taxonomic section.

Silurian ischnacanthiforms

The oldest known ischnacanthiform taxon erected for isolated jaw bones is probably *Xylacanthus kenstewarti* Hanke *et al.*, 2001b from the late Wenlock or early Ludlow of northern Canada. This species already shows the features characteristic of the main forms of Silurian–Middle Devonian ischnacanthiform jaw bones. Main cusps of the lateral teeth are perpendicular to the long axis of the jaw bone, and these cusps have small secondary cusps anterior and posterior; the bones have a concave base which straddles the jaw cartilage, and the posterior part of the jaw bone flares out laterally, and does not form part of the jaw articulation (Fig. 2A). Although Hanke *et al.* (2001b) described the lateral dentition row as comprising large monocuspid teeth separated by smaller conical denticles, the teeth appear to conform to the normal arrangement, as some of the large anterior teeth have the most posterior "denticles" running slightly medial to the "denticles" anterior to the following tooth, identical to the standard ischnacanthiform-type in which each tooth comprises a main cusp plus anterior and posterior secondary cusps. This typical structure was first recognised by Dean (1907) and reiterated by Goodrich (1909), Ørvig (1967) and Gross (1967).

Onchus graptolitarum is also of either Wenlock or Ludlow age; the type material is partly articulated fish with dentigerous jaw bones, fin spines, scales and scapulocoracoids in nodules from the "Liten beds", Bohemia (Czech Republic). Unfortunately, the specimens were

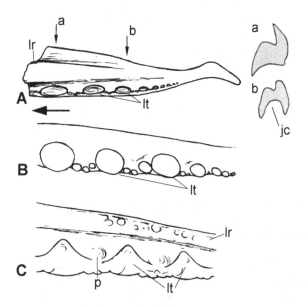

Fig. 2. General layout of Silurian–Early Devonian ischnacanthiform acanthodian dentigerous jaw bones. A: Occlusal view of jaw bone, with cross-sections. B: *Gomphonchus*-type jaw bone dentition. C: Poracanthodid-type jaw bone dentition. Anterior to left, lingual to top. jc, jaw cartilage concavity; lr, lingual ridge; lt, main cusps of lateral teeth; p, inter-tooth pit.

only poorly illustrated in the original description (Fritsch 1907), and have not been located.

All other acanthodian jaw bones known from the Silurian are isolated elements. The first to be described were from the Pridoli of the Ludlow bone bed in West Midlands, Britain, for which Agassiz erected *Plectrodus mirabilis* (in Murchison 1839). Priem (1911) also assigned jaw bones from the Upper Silurian of Portugal to this taxon. White (1961) erected a new species *Ischnacanthus kingi* for large jaws (about 9 cm long) from the Pridoli of Shropshire and Staffordshire, England. The main distinction recognised between jaw bones from this species and those of *Ischnacanthus gracilis* was the much larger size of the *Ischnacanthus kingi* jaws. Gross (1957, 1971) described two small (<1 cm long) jaw bone forms from the Late Silurian Beyrichienkalk of north Germany as *Nostolepis*- and *Gomphonchus*-type; these represent the two forms which are most characteristic of Silurian to late Early Devonian ischnacanthiforms. The "*Nostolepis*-type" resembles the poracanthodid-type as exemplified by *Zemlyacanthus menneri*, and is distinguished from the "*Gomphonchus*-type" (comparable with those of *Ischnacanthus gracilis*) by having lateral teeth in which the main cusps have a triangular parabasal section, and having denticles on a lingual ridge (Fig. 2B, C). These two categories of small jaw bones, which each appear to encompass a range of morphologies and presumably taxa, are commonly found in Late Silurian vertebrate assemblages throughout the Baltic region (Gross 1971; Vergoossen 1993), and are also reported from Greenland (Blom 1999), Arctic Canada (Burrow *et al.* 1999), Nevada (Burrow 2003) and Bolivia (Janvier & Suarez-Riglos 1986). Another less common form from the Baltic is the laterally compressed "*Acrodus*-type" (Gross 1947; Vergoossen 1993). The Australian record for Silurian ischnacanthiforms is meagre, with only small fragments of "*Gomphonchus*-type" jaw bones known from the Jack Formation (Ludlow), north Queensland (Burrow & Simpson 1995) and a shale deposit at Araluen, New South Wales (Burrow & Turner 2000).

Early Devonian ischnacanthiforms

Ischnacanthiform diversity increased markedly in the Lochkovian, and several species are based on articulated fish. *Ischnacanthus gracilis*, the type for the order, family and genus, is found at many Scottish localities in the Dundee Formation and the Lower Old Red Sandstone (Denison 1979). Based on the characters of the type species, *Ischnacanthus* jaw bones lack denticles on the lingual ridge, and the main cusps of the teeth are circular or oval in parabasal section (Fig. 2A). One figured specimen (NHM P29725), from the Clee district, England, which White (1961: pl. 42, fig. 3a, b) referred to

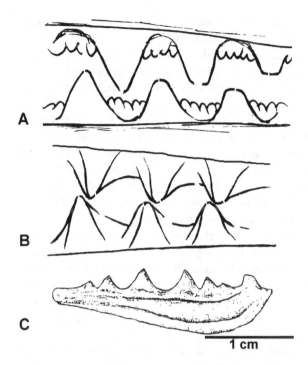

Fig. 3.   A, B: Two types of jaw occlusion in acanthodians with dentigerous jaw bones. A: Silurian–Early Devonian forms. B: Middle–Late Devonian forms. C: First figured specimen of *Acanthodopsis wardi* (Hancock & Atherton 1868, pl. 15, fig. 6), Upper Carboniferous Coal Measures, Newsham Colliery, northern England. Anterior to left.

Ischnacanthus wickhami, has a denticulated lingual ridge, so it should perhaps be referred to another genus.

As mentioned earlier, the jaw bones associated with the *Uraniacanthus spinosus* specimen on NHM P53032 described by Miles (1973) are probably referrable to *Ischnacanthus*. Although the dentition on these jaw bones was originally described as comprising monocuspid teeth of equal height, my impression from the plate figuring these elements (Miles 1973, pl. 12, fig. 1) is that the dentition layout has been misinterpreted. Some of the cusps in this photograph have been retouched, and it is this part of the jaw upon which the reconstruction is based (Miles 1973, fig. 17A). The more posterior teeth appear to show the typical *Ischnacanthus* pattern, although the main cusps have been worn down to stubs or are broken.

Articulated acanthodians are well represented in the Lower Devonian of Severnaya Zemlya in the Russian Arctic, and are being studied and described by Juozas Valiukevicius (Lithuanian Geological Institute). The first to be described (Valiukevicius 1992) was the poracanthodid *Zemlyacanthus (Poracanthodes) menneri*. All the new taxa with dentigerous jaw bones (*Acritolepis urvantsevi*, *A. ushakovi*, *Acanthospina irregulare*) described by Valiukevicius (2003) I would assign to the Ischnacanthiformes.

Xylacanthus grandis Ørvig, 1967, from the Pragian of Spitsbergen, was the largest ischnacanthiform, with jaws up to 35 cm long. These elements are structurally similar to those of *Xylacanthus kenstewarti* from Arctic Canada, with main tooth cusps having a circular parabasal section (characteristic of *Ischnacanthus*-type and "*Gomphonchus*-type" jaw bones), but *Xylacanthus grandis* differs from *Xylacanthus kenstewarti* in lacking denticles on the lingual ridge. *Xylacanthus minutis* Gagnier & Goujet, 1997, also from the Pragian of Spitsbergen, has jaws of a similar size to those of the older *Xylacanthus kenstewarti*, but the main teeth are triangular in parabasal section and the lingual ridge is denticulated. These "*Poracanthodes*-type" features indicate that *Xylacanthus minutis* should perhaps be referred to a different genus, as the other *Xylacanthus* species have a circular parabasal section.

Ischnacanthus? scheii Spjeldnaes, 1967 is a taxon based on associated jaw fragments, tooth whorl, fin spines and scales from a Lochkovian stratum of the Cape Phillips Formation, Ellesmere Island, Arctic Canada. The dentigerous jaw bone fragments have tricarinate teeth (i.e. triangular parabasal section) and are similar in size to those of *Xylacanthus minutis*. No secondary cusps are discernible between the lateral teeth on the jaw bone, the fin spines resemble those of poracanthodids (cf. Valiukevicius 1992, pl. 8, fig. 4), and the scales have ridged crowns, indicating that this taxon should not be assigned to the Ischnacanthidae.

The Lower Devonian of the western USA has yielded isolated ischnacanthiform jaw bones of at least two forms, one comparable with *Ischnacanthus gracilis* and the other a new species (Burrow 2002a). The type specimen of *Helenacanthus incurvus* Bryant, 1934, from the Beartooth Butte Formation of Wyoming, was originally described as an acanthodian spine, but Denison (1979) reassessed it as an ischnacanthiform dentigerous jaw bone. However, by comparison with the long, shallow infragnathals of *Actinolepis spinosa* Mark-Kurik, 1985 and an indeterminate actinolepid from Utah (Denison 1958, fig. 101F), it is probably an actinolepid placoderm infragnathal.

Although East Gondwana has no articulated ischnacanthiform acanthodians from the Lower Devonian, fragments of small jaw bones (estimated total length <0.5 cm) and tooth whorls, as well as ischnacanthiform scales, are common Early Devonian microremains from eastern Australia (Burrow 1995, 1997, 2002b). The East Gondwanan acanthodian fauna becomes more endemic during this timespan. Gnathostome macroremains are common in the late Pragian–early Emsian limestones of the Taemas-Wee Jasper region near Canberra, ACT, and include small- to medium-sized (<4 cm long) dentigerous jaw bones. The first of these to be formally described was *Taemasacanthus erroli* Long, 1986; several new taxa have recently been added to the list (Lindley 2000, 2002), although the status of some of these is debatable (Burrow 2002b). Nearly all of these Australian late Early Devonian elements conform to the general structure exhibited by

Taemasacanthus erroli (Burrow 2002b): they are characterised by robust teeth forming both lateral and lingual rows, and by the secondary cusps on the lateral tooth row being reduced to denticulations on a flange. *Rockycampacanthus milesi* Long, 1986 is a distinctive taxon from the Emsian Murrindal Limestone, Victoria, distinguished by its double row of small secondary cusps posteromedial to the main lateral cusps (Long 1986, fig. 2A, B). The only other East Gondwanan terrane which has yielded Lower Devonian ischnacanthiform jaw bones is south China: *Youngacanthus gracilis* Wang, 1984 was erected for fragments of "*Poracanthodes*-type" dentigerous jaw bones, estimated total length about 1 cm, from the Xitun Member (Lochkovian), Cuifengshan Formation.

Having eliminated *Helenacanthus* and *Uraniacanthus* from the ischnacanthiforms, a review of the Silurian to Early Devonian taxa shows that nearly all share a noteworthy character: when viewed from a lateral or medial perspective, the jaws have upright main tooth cusps separated by deep pits for the occluding teeth, and the teeth of the upper and lower jaws come together in a cog-like action (Fig. 3A).

Middle–Upper Devonian ischnacanthiforms

Many Middle and Upper Devonian isolated dentigerous jaw bones have been assigned to the genus *Atopacanthus*. The type species is *Atopacanthus dentatus* Hussakof & Bryant, 1918, and the type specimen E2496 is a 2.5 cm long jaw fragment collected by Bryant from the forks of Eighteen Mile Creek, near Hamburg, New York; the type stratum is the Frasnian Rhinestreet shale (Portage). This taxon is characterised by monocuspid teeth forming the main lateral row, and has been redescribed recently (Burrow in press). The oldest Middle Devonian taxon is *Atopacanthus peculiaris* Hussakof, 1913, based on jaw fragments from the Eifelian Onondaga Limestone of New York. The elements were originally thought to be fin spine fragments and were assigned to *Apateacanthus*: the type specimen of *Apateacanthus* is a fin spine fragment from the Upper Devonian of New York. Denison (1979, p. 38, fig. 26G) included the type species *Apateacanthus vetustus* (Clarke, 1885) in the ischnacanthiforms, but as the genotype is a fin spine and not a jaw bone as described by Denison, there is no evidence for assigning *Apateacanthus* to the Ischnacanthiformes (Burrow in press, fig. 5A). The type material of *Atopacanthus peculiaris*, however, includes a weathered full-length jaw bone plus two other jaw fragments in matrix (Hussakof 1913, pl. 47, figs. 4–6). These specimens are poorly preserved, weathered bone lacking the outer surface, so that the shape of neither the lateral nor the medial surfaces is determinable. On cotype NYSM P10330 (Burrow in press, fig. 5B), the medial surface is buried, and the presence or absence of a lingual ridge is not determinable; the posterior part of the bone angles out at about 30° to the long axis of the jaw.

Atopacanthus peculiaris marks the transition between older and younger ischnacanthiform jaw bones; unlike most of the older jaw bones, jaws of this taxon lack secondary cusps on the lateral teeth. Such cusps are also lacking in other ischnacanthiform jaw bones of Eifelian age, e.g. *Atopacanthus? ambrockensis* Otto, 1999 from the Brandenburg Group, Germany, and an indeterminate ischnacanthiform from the Khush-Yeilagh Formation, Iran (Blieck *et al.* 1980).

All ischnacanthiforms known from Middle–Upper Devonian deposits worldwide have monocuspid lateral teeth, and have been assigned (sometimes tentatively) to either *Atopacanthus* or *Persacanthus* Janvier, 1977. Long *et al.* (in press) are revising these taxa. The principal functional difference between the dentition of pre-Middle and post-Early Devonian ischnacanthiforms is reflected in the shape of the main lateral teeth. Ørvig (1973, p. 127) commented that he had seen no wear marks on *Atopacanthus* jaw bones; presumably he was referring to blunting of tooth apices, rather than microwear on the sides of the teeth. The change in tooth shape and wear was probably related to a change from a vertical cog-like biting motion to a shearing action that tended to sharpen rather than blunt the teeth. Jessen (1973) hypothesised that the teeth of opposing jaws on his *Atopacanthus?* sp. fish from the Upper Plattenkalk of Bergisch-Gladbach occluded by sliding down along the posterior surface of the opposing tooth into the inter-tooth pit, guided by the posterior flanges (Fig. 3B). The change in biting strategy may be correlated with the increase in the angle at which the posterior end of the jaw bone diverges from the main jaw axis, which is undoubtedly related to the angle that the articulation makes with the jaw axis. In many Middle–Late Devonian taxa this angle approaches 90°; presumably, forward/backward sliding of opposing jaws was thus prevented, while some sideways movement was still possible.

Carboniferous "toothed" acanthodians

In the Carboniferous, the only acanthodian dentigerous jaw bones are of the *Acanthodopsis* form. The type specimen of *Acanthodopsis wardi* Hancock & Atthey, 1868 is a head plus pectoral region with jaws about 4 cm long from Newsham Colliery, in the Upper Carboniferous Coal Measures of Britain. *Acanthodopsis egertoni* Hancock & Atthey 1868 was erected for much larger fish (also from Newsham Colliery) estimated to have been about 75 cm long, with large pectoral spines, smooth-crowned scales, and jaws with large, laterally compressed, finely striated, triangular "teeth" forming a single row on each ramus. However, Woodward (1891) and Denison (1979) regarded this taxon as a junior synonym of *A. wardi*. Most

other specimens assigned to *A. wardi* are isolated dentigerous upper and lower jaws about 2.5 cm long (e.g. Miles 1966, fig. 12B; Fig. 3C). Over the years, the taxon has been assigned either to the Ischnacanthiformes (based on the dentigerous jaw bones) or to the Acanthodiformes (based on the acanthodiform-type double mandibular joint). *Marsdenius* Wellburn, 1902, in the absence of specimens or illustrations of this taxon (see above), is not considered in determining the relationships of *Acanthodopsis*.

Jaw structure and articulation

An understanding of the jaw structure and articulation in Early Devonian (and older) ischnacanthiforms gives a foundation for comparisons between the different forms of dentigerous jaw bones found in acanthodians. The dentigerous jaw bones which characterise the ischnacanthiforms are of dermal origin. Ørvig (1973) clarified this point in his response to Miles' (1965) description of the teeth as being ankylosed to ossified parts of the jaw cartilages, rather than being part of marginal dermal elements. Certainly, the jaw bones were in direct contact with jaw cartilages, and in rare specimens with both the mineralised jaw cartilage and the jaw bone preserved; the boundary between the two tissues is distinguishable (e.g. Valiukevicius 1992, pl. 2, fig. 1; Hanke *et al.* 2001b, figs. 3C, 4C). Isolated jaw bones usually have a relatively smooth basal surface, forming the concavity for the jaw cartilages (e.g. Lindley 2000, fig. 9D, 2002, figs. 5D, F, 7D, E).

The articulation surfaces of the jaws are on the cartilages, not the jaw bones, and there is rarely any direct evidence of the provenance – upper or lower jaw – of most isolated jaw bones. In rare examples, these surfaces on the palatoquadrate and Meckel's cartilage have been perichondrally ossified or calcified; some isolated jaws show this type of preservation and can thus be categorised as from the upper or lower jaw (e.g. Long 1986, fig. 3C, E; Lindley 2002, fig. 8A, D). Articulated fish, for example *Zemlyacanthus menneri* (Valiukevicius 1992, pl. 2, fig. 1) and *Ischnacanthus gracilis* (Fig. 4B, C), often have the whole jaw cartilage mineralised. Thus, it is possible to reconstruct the jaw articulation, for example based on the *Ischnacanthus gracilis* specimens from the Forfarshire quarries in Scotland, even though the specimens are usually squashed and slightly disarticulated.

Ischnacanthus gracilis has a simple mandibular joint, with the posterior end of Meckel's cartilage forming an oblique posterodorsal process (*sensu* Long 1986) or condyle (Fig. 4A) which articulated with an oblique socket/cotylus on the posterior-most corner of the palatoquadrate cartilage (Fig. 4C; Watson 1937). The jaw joints in some ischnacanthiform acanthodians from the Lochkovian of northwest Canada (Gagnier & Wilson 1995) are comparable with that of *Ischnacanthus gracilis*. I consider the taemasacanthids to have had the same type

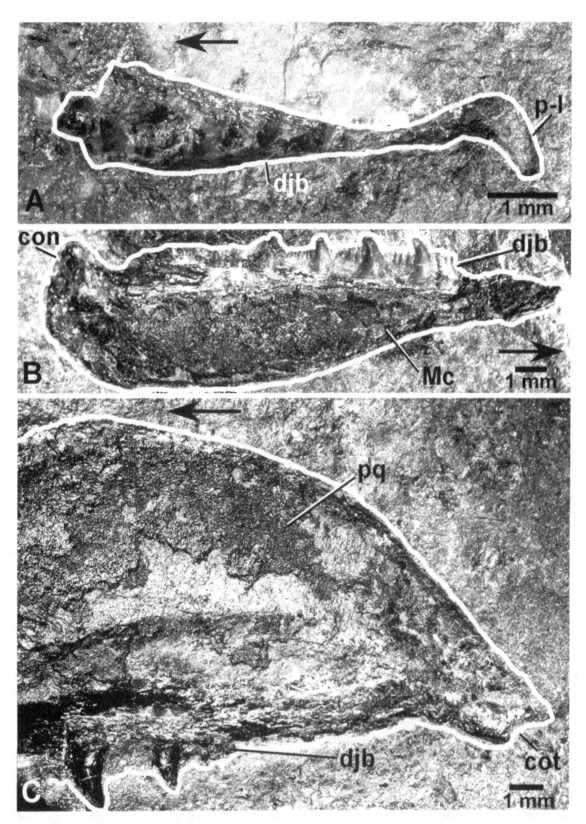

Fig. 4. Ischnacanthus gracilis jaws; specimens from Lochkovian Dundee Formation, Tillywhandland Quarry, Forfarshire, Scotland (elements outlined in Adobe Photoshop). A: Occlusal view of dorsoventrally squashed jaw (QM CB17b). B: Medial view of lower jaw with mineralised Meckel's cartilage, and dentigerous jaw bone (QM CB22a). C: Lateral view of upper jaw with mineralised palatoquadrate, and dentigerous jaw bone (QM CB16a). con, condyle; cot, cotylus; djb, dentigerous jaw bone; Mc, Meckel's cartilage; p-l, posterolateral end of dentigerous jaw bone; pq, palatoquadrate.

of joint. Long (1986, fig. 3A–E) showed an articular foramen and an apparently concave articular surface on a
lower jaw of *Taemasacanthus erroli*, but I believe this
shape was caused by the collapse of the cartilaginous core
within the perichondral ossification and that, in life, the
structure was condylar. The paratype specimen ANU
60109 of *Taemasacanthus cooradigbeensis* Lindley (2002,
fig. 8A–E) is also a lower jaw element in which the articular condyle of Meckel's cartilage has collapsed. Following
this interpretation, all ischnacanthiforms known from
articulated fish have the same simple type of jaw joint.

Another topic to be considered is the growth of the
ischnacanthiform jaw bones. Tooth replacement in
acanthodians was compared with that of shark teeth by
early workers (e.g. Dean 1907). Because the teeth on the
dermal jaw bones of ischnacanthiforms are not replaced
and remain ankylosed to the basal bone once formed,
Gross (1967) suggested that the jaw bones were shed at
intervals and replaced by newly formed ones. In contrast,
Ørvig (1973, p. 129) promoted the current theory of tooth
addition at the anterior end of the jaw bone: he noted
that even on completely preserved jaw bones, the bone
tissue often terminated anteriorly in an irregular manner,
indicating "an ever-present zone of growth where, during
intervals of heightened scleroblastic activity, new bone
tissue could form in direct continuity with the earlier
existing hard tissue of this kind...". Lindley (2000) elabo-
rated on this concept, showing that in some taemasacanthids (and perhaps in all ischnacanthiforms) the basal
bone grew before the new teeth were added. Growth
of the dentigerous jaw bones differed to growth of other
teeth and tooth whorls in the mouth and pharyngeal
region of ischnacanthiforms and other acanthodians. In a
broad sense, the dermal jaw bone can be compared with a
fin spine, with growth occurring at the anterior part of
the jaw bone analogous to the growth of the proximal part
of the fin spine. Possibly, some forms of oropharyngeal
tooth whorls and teeth, while having an ectodermal
origin (Goodrich 1930; Mallatt 1984; *contra* Smith &
Coates 1998) and having scale-like progenitors, could
have arisen first in the pharynx (Smith & Coates 1998),
but most acanthodian dental elements show similarities
with extra-oral dermal structures.

Systematic Palaeontology

Class Acanthodii Owen, 1846

Order Acanthodiformes Berg, 1940

Family Acanthodidae Huxley, 1861

Genus *Acanthodopsis* Hancock & Atthey, 1868

Type species. – *Acanthodopsis wardi* Hancock & Atthey,
1868.

Revised diagnosis. – Acanthodid fish having the whole of
the palatoquadrate and Meckel's cartilage perichondrally
ossified, with the ossification over Meckel's cartilage
continuing down to a separate, ventral mandibular splint
bone; at least two ossification centres in the upper jaw;
articulation and general form of the jaws (except the
"teeth") typically acanthodiform with a double jaw
articulation; mandibular splint and/or perichondral ossification forms a post-articular process; dentition comprises a single row of up to 11 monocuspid, triangular,
vertically striated, tooth-like structures on each ramus
with the largest "teeth" in the middle of the row; fin spines
smooth except for a groove near the anterior margin;
robust scapulocoracoid with ossified suprascapula and
scapula, and cylindrical procoracoid process.

Remarks. – Long (1986), following most early workers,
advocated an acanthodiform connection for *Acanthodopsis* based on the type of jaw joint: i.e. a "double" articulation with pre-glenoid and post-glenoid processes on
the lower jaw, and pre-articular and articular processes
on the upper jaw (cf. Long 1986, fig. 6). *Acanthodopsis
wardi* has a post-articulation process on the lower jaw
(Hancock & Atthey 1868, pl. 15, fig. 6). The scales, fin
spines and scapulocoracoid are also comparable with
those of *Acanthodes* spp.; Hancock & Atthey (1868)
referred *Acanthodes wardi* Egerton, 1866 to *Acanthodopsis*, but as *Acanthodes wardi* is an articulated acanthodiform fish without teeth, Denison (1979) excluded it
from *Acanthodopsis*.

Acanthodopsis russelli sp. nov.
Figs. 5–9

Synonymy. –

2002 "*Acanthodopsis*-type dentigerous jaw bone" –
　　Burrow & Turner, p. 193.

Etymology. – For the Russell family, owners of Plain Creek
station.

Type specimens. – Holotype: right lower jaw with mandibular splint (QM51266; Figs. 5A, B, 6, 9A). Paratypes:
right palatoquadrate (QM51267; Figs. 5C, D, 9A); left
lower jaw on QM51268 (Fig. 7A, B); partial right lower
jaw on QM51269 (Fig. 7C).

Other material. – Four right lower jaws on QM51271–
51274; six left lower jaws on QM51275–51280; one right
palatoquadrate on QM51281; two left palatoquadrates
on QM51282 (Fig. 7D) and QM51283; jaw and spine
fragments on QM51288–51289; four scapulocoracoids
on QM51276 (Fig. 8A–C), QM51284–51286; right

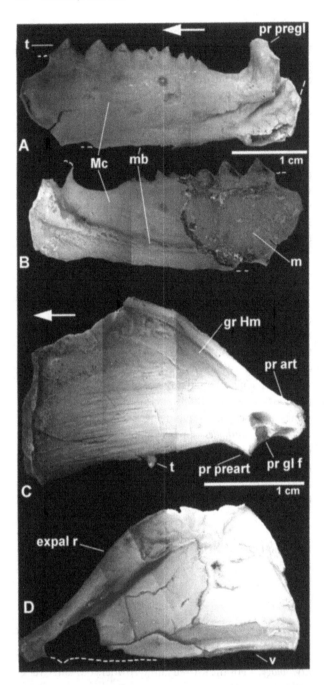

Fig. 5. *Acanthodopsis russelli* sp. nov. from locality QML1156, Drummond Basin, Queensland, Australia; Bulliwallah Formation (mid Viséan). A, B: Cast of holotype QM51266, right lower jaw comprising Meckel's cartilage and mandibular splint; anterior end missing (A, medial view; B, lateral view). C, D: Cast of paratype QM51267, right palatoquadrate (C, medial view; D, lateral view). expal r, extrapalatoquadrate ridge; gr Hm, groove for hyomandibular; m, mould of preserved jaw tissue; mb, mandibular splint; Mc, Meckel's cartilage; pr art, articular process; pr gl f, pre-glenoid fossa of palatoquadrate; pr preart, pre-articular process; pr pregl, pre-glenoid process; t, tooth; v, ventral shelf of palatoquadrate. Arrow is anteriad. All Exaflex casts whitened with magnesium oxide, photographed with an Olympus DP12 imaging system and SZ40 microscope, and compiled using Adobe Photoshop.

metapterygoid on QM51270; fin spines, scales, possible circumorbital/sclerotic bones on QM51287. Most specimens are preserved as impressions; any friable bone remaining was removed with dilute hydrochloric acid. The terminal parts of elements which were protruding from the nodules and exposed to weathering were usually lithified as siderite, and could not be dissolved, nor prepared away from the soft matrix.

Type locality and horizon. – Bulliwallah Formation, mid Viséan, Holkerian, Early Carboniferous; QML1156, "the Hut", Plain Creek, 21°33.04'S 146°29.95'E, northwest of Belyando Crossing, central north Queensland. The only known occurrence is at the type locality, where specimens were collected by the author and S. Turner, 2001–2002.

Diagnosis. – Acanthodopsid acanthodid fish; medial surface of lower jaw bones convex; some lower jaws with approximately nine "teeth" on the occlusal surface: smallest "teeth" posterior and largest "teeth" anterior, with weakly striated medial faces and smooth lateral faces; straight mandibular splint; upper jaws with few or no "teeth"; "teeth", when present, with a circular parabasal section; metapterygoid separately ossified to other parts of palatoquadrate, with thickened rim bordering posterodorsal quadrant of orbit.

Remarks. – Despite the variation in "dentition" on the jaws, all acanthodid elements from the QML1156 locality are provisionally assigned to *Acanthodopsis russelli* sp. nov. The presence or absence and extent of "teeth" on the jaws seems to be the only variable character noted from any of the elements, and that character appears to be independent of the size of the jaws. The overall shape of the lower jaws of *A. russelli* sp. nov. closely resembles that of another species, *Acanthodopsis microdon* Traquair, 1894, which was erected for small jaw bones (about 2 cm long) from the Upper Carboniferous Woodhouse Coal, Cheadle Coalfield, England. The Queensland fossils were all found in a small area and are probably the remains of the same community.

Description. – All specimens were fossilised as disarticulated, uncompressed, three-dimensional elements in a homogeneous, unlaminated silty deposit. The nodules in which nearly all the specimens were found are the secondary product of surface weathering of the siltstone, which is often barely lithified; presumably the calcium phosphate of the bones and coprolitic matter acted (to a limited degree in this instance) as a permineralising agent.

Lower jaws

None of these elements has their whole length preserved; the specimens are from 14 to 58 mm long. The holotype QM51266 (Fig. 5A, B) is 36 mm long and 12 mm high

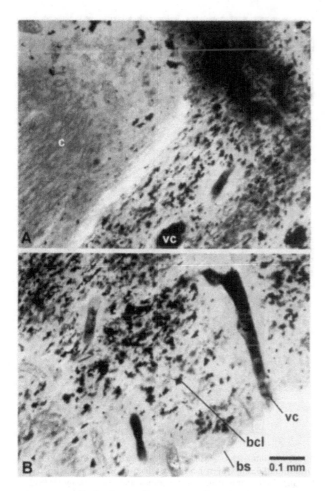

Fig. 6. *Acanthodopsis russelli* sp. nov. from locality QML1156, Drummond Basin, Queensland, Australia; Bulliwallah Formation (mid Viséan). Ground thin section through mandibular splint of the holotype, QM51266. A: Denser inner layer (?cartilage) and outer layer of vascular bone. B: Outer layer with ?bone cell lacunae and vascular canals. bcl, bone cell lacuna; bs, surface of bone; c, ?cartilage core; vc, vascular canal.

at both ends (6 mm posteriorly and 10 mm anteriorly without the mandibular splint). The anterior part of the jaw is missing. Before preparation by clearing with hydrochloric acid, the whole of Meckel's cartilage was covered with a thin perichondral ossification. The mandibular splint was 2 mm thick and formed of cellular bone pierced by vascular canals in the outer layer, and a denser inner layer (Fig. 6A, B). The medial face of Meckel's cartilage is convex, and the lateral face slightly concave, with the ventral limit of the fossa for the adductor musculature delimited by the upper edge of the mandibular splint (Fig. 5A, B). The ventral surface of the mandibular splint is relatively straight from the anterior end, as preserved, to the posterior limit, where it forms a bulge in the jaw outline. The jaw ossification then curves upwards posteriad to form a post-glenoid process behind the main jaw joint (the process was replaced by siderite, and is thus not visible on the cast). This process is separated from the

bulbous pre-glenoid process by a cup-like cavity; both processes are angled posteromedially. The jaw narrows slightly in front of the articulation areas, then deepens along the occlusal surface, which bears a row of monocuspid "teeth". As on the rest of the jaw, the bone covering the "teeth" is thin, and the inner tissue is only preserved as a brown, porous mass; the "teeth" have weak wavy striations on their medial faces. The three posterior-most "teeth" are all small, the fourth is notably larger, the fifth and sixth are shorter than the fourth, with the seventh, eighth, and ninth larger, and progressively increasing in height (Fig. 5A, B). The front end of the jaw is not preserved, and so the size of more anterior teeth is not known. On most of the other lower jaws collected, the occlusal surface is obscured by other remains, so that the presence or absence of "teeth" is not clearly determinable. However, the bases of about seven "teeth" are visible on QM51268 (Fig. 7A, B), one of the smallest jaws at 26 mm long. The dimensions of the "tooth" bases indicate that its "teeth" were largest in the middle of the jaw. On QM51277 (Fig. 7C), the posterior half of a right lower jaw, the upper edge is reasonably well preserved, but bears no "teeth". Most specimens appear to be only the posterior, articular ossifications of the lower jaws, without the anterior, mentomandibular ossification. Except for the dentition, all the lower jaws share the same structure. The jaw tissue has been preserved in QM51275; unfortunately, this specimen is only the posterior part of the jaw and no "teeth" have been detected. The counterpart is an impression of the medial surface, which was cleared with hydrochloric acid and shows that although very thin, the perichondral bone was multilayered.

Upper jaws

Paratype QM51267 (Fig. 5C, D) is a right palatoquadrate 28 mm long and 15 mm high at the anterior limit of the hyomandibular groove. As on the lower jaws, thin perichondral ossification covered the whole surface. The articular surface has prominent articular and pre-articular processes separated by a deep cotylus, which is the fossa for the pre-glenoid process of the lower jaw. A smooth, rounded rim runs along the dorsal edge of the medial surface of the palatoquadrate, paralleled by a deep groove for the hyomandibula (Fig. 5C). The specimen bears just one small tooth near the middle of the ventral/occlusal shelf. The laterodorsal margin of the palatoquadrate shows a well-developed extrapalatoquadrate ridge which is separated from the ventral shelf by a deep, extensive fossa for the adductor musculature. Casts of the holotype lower jaw (QM51266) and paratype palatoquadrate (QM51267) articulate perfectly (Fig. 9A). None of the other palatoquadrates has "teeth" visible (Fig. 7D). QM51281 shows the vertical cross-sectional shape of the palatoquadrate (Fig. 9B). An isolated metapterygoid (QM51270; Fig. 7E) has an anterodorsal, bowl-shaped

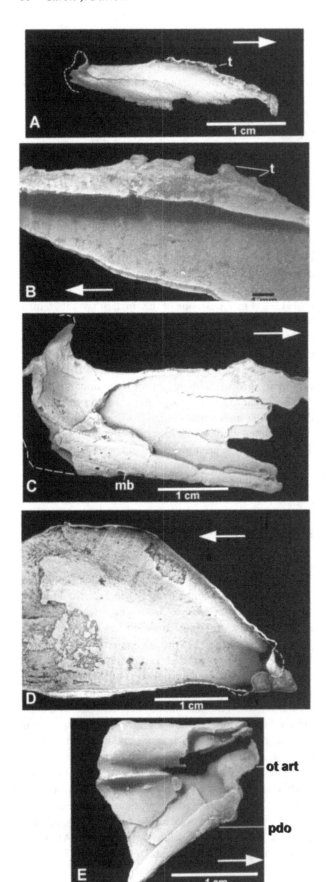

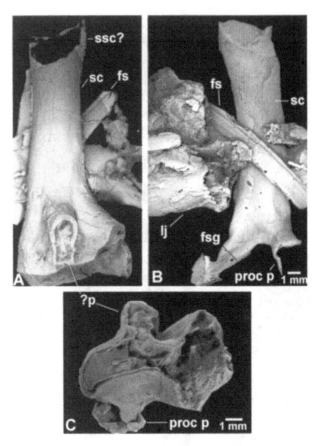

Fig. 8. *Acanthodopsis russelli* sp. nov. from locality QML1156, Drummond Basin, Queensland, Australia; Bulliwallah Formation (mid Viséan). Right scapulocoracoid, fin spines, and lower jaw on QM51276 (A, posteromedial view; B, anterolateral view; C, basal view). fs, fin spine; fsg, groove for pectoral fin spine; lj, lower jaw; proc p, procoracoid process; ?p, indeterminate process; sc, scapulocoracoid; ssc, suprascapula. All Exaflex casts whitened with magnesium oxide, and photographed with an Olympus DP12 imaging system and SZ40 microscope.

depression/cotylus which articulated with the otic condyle of the neurocranium, and a thickened rim bordering the posterodorsal quadrant of the orbit.

Scapulocoracoids

The robust dorsal ends of the scapulocoracoids have a circular cross-section; the extremities have been exposed and are sideritic, and thus not removed by hydrochloric acid. QM51276 (Fig. 8A–C) is 22 mm high, with a top diameter of 5 mm. QM51284 is 20 mm high, QM51285

Fig. 7. *Acanthodopsis russelli* sp. nov. Jaw elements from locality QML1156, Drummond Basin, Queensland, Australia; Bulliwallah Formation (mid Viséan). A, B: Cast of paratype QM51268, dentigerous left lower jaw. A: Medial and B: lateral close-up of upper surface, partial cast. C: Cast of posterior part of right lower jaw QM51269, lateral view. D: Cast of left palatoquadrate QM51282 (CB02-27), lateral view. E: Cast of right metapterygoid QM51270, lateral view. mb, mandibular splint; ot art, concavity for otic articulation; pdo, rim delimiting posterodorsal edge of orbit; t, teeth. Arrow is anteriad. All Exaflex casts whitened with magnesium oxide, and photographed with an Olympus DP12 imaging system and SZ40 microscope.

is incomplete (only upper 10 mm of shaft preserved), and QM51286 is 13 mm high. All have a top diameter of 4 mm. The scapular shaft narrows slightly then expands ventrally to form a posterior flange, an anterolateral, cylindrical procoracoid process, and a ventromedial cylindrical process. The function of the latter is unknown, but it is in the same position as the subscapula fossa on *Acanthodes bronni* (Miles 1973, text-fig. 19). The coracoid part of the endoskeletal shoulder girdle does not appear to have been ossified, or at least not co-ossified with the scapula. These scapulocoracoids (Fig. 9D), and the many fin spines and scales preserved in the nodules, closely resemble those of *Acanthodes wardi* (Miles 1973, text-fig. 21) from the Late Carboniferous Coal Measures of Longton, England (Fig. 9E), and also those of an

acanthodid (jaws illustrated by Long 1986) from the Early Carboniferous Raymond Formation of central Queensland (Fig. 9C).

Discussion

The new vertebrate taxa which have been described in recent years from the Lower Carboniferous of central Queensland provide evidence for a close similarity between these faunas, the Early Carboniferous faunas of the Midland Valley, Scotland, and the Late Carboniferous (Westphalian A and B) faunas of the British Coal Measures. Except for the presence of "teeth" on some of the jaws, all of the new acanthodid material (jaws, scapulocoracoids, fin spines, scales) from "the Hut" locality QML1156 is structurally comparable with elements of the slightly older Raymond Formation acanthodid species (Fig. 9C, D). However, the size of the elements indicates that average fish from QML1156 were about twice the size of the Raymond Formation species. It hardly seems coincidental that a similar situation is repeated in the younger British faunas, where the two types of fish (*Acanthodopsis wardi* and *Acanthodes wardi*) have been assigned to different species (and orders!: Miles 1966) despite the ossified elements differing only in the presence or absence of "teeth" on the jaws. Given this similarity, the original 19th century assessment (Hancock & Atthey 1868) of the British taxa *Acanthodes wardi* and *Acanthodopsis wardi* as dimorphic varieties of the same species is understandable. My investigations of catalogue lists from the American Museum of Natural History, Hancock Museum, Natural History Museum (London), National Museum of Scotland, Manchester Museum, Museum of Comparative Zoology (Harvard), and University Museum of Zoology (Cambridge) indicate that dentigerous jaws assigned to *Acanthodopsis*, and articulated specimens (with heads) of *Acanthodes wardi*, have not been collected from the same locality. However, some confusion is generated by the National Museum of Scotland having synonymised *Acanthodes wardi* with *Acanthodopsis wardi*, and isolated spines of identical morphology appear to have been assigned indiscriminately to either *Acanthodes* or *Acanthodopsis*. A re-examination of the catalogued specimens is needed to clarify the distribution of the two taxa.

Many workers have noted the difference in structure of the dentigerous jaw bones of *Acanthodopsis* compared with those of typical ischnacanthiform acanthodians (Miles 1973; Ørvig 1973). The new material provides further evidence that the "teeth" of *Acanthodopsis* are not real teeth, and are convergent rather than homologous with those of ischnacanthiforms. All other structures (fin spines, scapulocoracoids, jaw cartilage articulation,

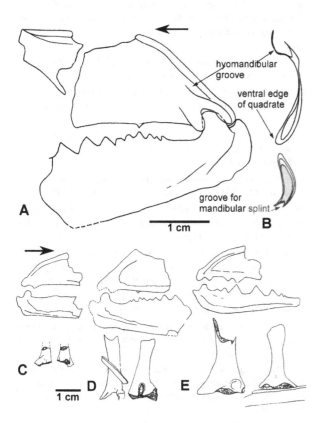

Fig. 9. Comparison of some Carboniferous acanthodid jaws and shoulder girdles. A, B: Lower Carboniferous (?mid Viséan; Bulliwallah Formation) *Acanthodopsis russelli* sp. nov. jaws from the northern Drummond Basin, central Queensland, Australia. A: Jaws in articulation, restored in medial view, based on QM51266, 51267, and 51270. B: Vertical cross-sections of the palatoquadrate (QM51281) and Meckel's cartilage (QM51278), medial surfaces to right. C–E: Three selected Carboniferous acanthodids; lateral views of right jaws (above), and lateral and medial views of right scapulocoracoids (elements figured at approximately the same scale). C: Lower Carboniferous (?early Viséan; Raymond Formation) acanthodid jaws (modified from Long 1986, fig. 6A, B) and base of scapulocoracoid (QM51290, middle Drummond Basin, central Queensland, Australia). D: *Acanthodopsis russelli* sp. nov. E: *Acanthodopsis wardi* (Upper Carboniferous, British Coal Measures), jaws modified from Miles (1966, fig. 12A) and Hancock & Atthey (1868, pl. 2, fig. 6); scapulocoracoids and fin spine modified from Miles (1973, fig. 21A, B).

general shape, and ossification centres) are comparable with those of other, toothless acanthodids.

A possible explanation for the toothless and "toothed" variants is that the latter condition was related to mating, as exemplified by the elaboration of jaws in breeding male salmon; however, but the presence of "teeth" on both the smallest and largest of the lower jaws suggests the condition is not an adult versus juvenile difference. Alternatively, the "toothed" *Acanthodopsis* morphotypes could represent an ancient instance of a recessive state being maintained in a population: a mutation (namely tooth-like cones on the jaws) arose in a recessive state; the mutation (possibly sex-linked) was passed on to further generations, without phenotypic expression, until the population had enough heterozygotes so that some mated to produce homozygotes for the mutated recessive allele (cf. Schwartz 1999). In this scenario, even on some homozygous individuals the mutation could only be partially expressed, a possible example being the paratype palatoquadrate of *A. russelli* sp. nov. with only one "tooth" (Fig. 5C). Three windows on this phylogenetic development are opened by the three different-aged occurrences of similar skeletal elements: the Raymond Formation "pure" acanthodid species, without "teeth" on the jaws; the Bulliwallah Formation species, with partial expression of the "teeth" on some individuals; and British Coal Measures *Acanthodopsis wardi*, with "teeth" being the norm (Fig. 9C–E). The appearance of "mutant" homozygotes in a population has been touted as the source of new species (Schwartz 1999). Whereas in *Acanthodopsis russelli* sp. nov., the "toothless" and "toothed" phenotypes would have been maintained, the British taxa *Acanthodes wardi* and *Acanthodopsis wardi* might represent the products of such speciation.

Acknowledgements

I am grateful for the support of an ARC Postdoctoral Research Fellowship for 2002–2004. I also wish to thank the Russell family and workers of Plain Creek station for their hospitality and encouragement, Dr Susan Turner for inviting me to participate in the field work and study of the collection, Dr Bruce Burrow for keeping the fires alight (during field work), and reviewers Drs Gavin Hanke and Jo Vergoossen for their helpful and insightful comments on the manuscript. This is a contribution to IGCP 491: Middle Palaeozoic Vertebrate Biogeography, Palaeogeography, and Climate.

References

Berg, L.S. 1940: Classification of fishes, both Recent and fossil. *Trudy Zoologicheskogo Instituta. Akademiya Nauk SSSR. Leningrad 5*, 87–517.

Bernacsek, G.M. & Dineley, D.L. 1977: New acanthodians from the Delorme Formation (Lower Devonian) of N.W.T., Canada. *Palaeontographica A 158*, 1–25.

Blieck, A., Golshani, F., Goujet, D., Hamdi, B., Janvier, P., Mark-Kurik, E. & Martin, M. 1980: A new vertebrate locality in the Eifelian of the Khush-Yeilagh Formation, Eastern Alborz, Iran. *Palaeovertebrata 9*, 133–154.

Blom, H. 1999: Vertebrate remains from Upper Silurian–Lower Devonian beds of Hall Land, North Greenland. *Geology of Greenland Survey Bulletin 182*, 1–80.

Brotzen, F. 1934a: Erster Nachweis von Unterdevon im Ostseegebiete durch Konglomeratgeschiebe mit Fischresten. II Teil (Paläontologie). *Zeitschrift für Geschiebeforschung 10*, 1–65.

Brotzen, F. 1934b: Die Morphologie und Histologie der Proostea (Acanthodiden) Schuppen. *Arkiv für Zoologi Band 26A*, 1–27.

Bryant, W.L. 1934: The fish fauna of Beartooth Butte, Wyoming. Parts II and III. *Proceedings of the American Philosophical Society 73*, 127–162.

Burrow, C.J. 1995: Acanthodian dental elements from the Trundle Beds (Lower Devonian) of New South Wales. *Records of the Western Australian Museum 17*, 331–341.

Burrow, C.J. 1996: Taphonomic study of acanthodians from the Devonian Bunga Beds (Late Givetian/Early Frasnian) of New South Wales. *Historical Biology 11*, 213–228.

Burrow, C.J. 1997: Microvertebrate assemblages from the Lower Devonian (*pesavis/sulcatus* zones) of central New South Wales, Australia. *Modern Geology 21*, 43–77.

Burrow, C.J. 2002a: Evolution and relationships of acanthodians with dentigerous jaw bones. *Geological Society of Australia, Abstracts 68*, 26–27.

Burrow, C.J. 2002b: Lower Devonian acanthodian biostratigraphy of south-eastern Australia. *Memoirs of the Association of Australasian Palaeontologists 27*, 75–137.

Burrow, C.J. 2003: Poracanthodid acanthodian from the Upper Silurian (Pridoli) of Nevada. *Journal of Vertebrate Paleontology 23*, 489–493.

Burrow, C.J. in press: A redescription of *Atopacanthus dentatus* Hussakof and Bryant 1918 (Acanthodii, Ischnacanthidae). *Journal of Vertebrate Paleontology*. In press.

Burrow, C.J. & Simpson, A.J. 1995: A new ischnacanthid acanthodian from the Late Silurian (Ludlow, *ploeckensis* Zone) Jack Formation, north Queensland. *Memoirs of the Queensland Museum 38*, 383–395.

Burrow, C.J. & Turner, S. 2000: Silurian vertebrates from Australia. *Courier Forschungsinstitut Senckenberg 223*, 169–174.

Burrow, C.J. & Turner, S. 2002: Unusual preservation of vertebrate remains from the Carboniferous of north Queensland. *Geological Society of Australia, Abstracts 68*, 193.

Burrow, C.J., Vergoossen, J.M.J., Turner, S., Uyeno, T. & Thorsteinsson, R. 1999: Microvertebrate assemblages from the Late Silurian of Cornwallis Island, Arctic Canada. *Canadian Journal of Earth Sciences 36*, 349–361.

Clarke, J.M. 1885: On the higher Devonian faunas of Ontario County, New York. *Bulletin of the U.S. Geological Survey 16*, 1–80.

Dean, B. 1907: Notes on acanthodian sharks. *American Journal of Anatomy 7*, 209–222.

Denison, R.H. 1958: Early Devonian fishes from Utah. 3. Arthrodira. *Fieldiana Geology 11*, 459–551.

Denison, R.H. 1979: *Acanthodii. In* Schultze, H.-P. (ed.): *Handbook of Paleoichthyology*, Vol. 5. Gustav Fischer, Stuttgart.

Egerton, P.G. 1861: British fossils. *Memoirs of the Geological Survey of the United Kingdom (British Organic Remains) Dec. X*, 51–75.

Egerton, P.G. 1866: On a new species of *Acanthodes* from the Coal-shales of Longton. *Quarterly Journal of the Geological Society 22*, 468–470.

Fritsch, A. 1907: *Miscelanea palaeontologica. I. Palaeozoica.* Fr. Řivnác, Prague.

Gagnier, P.-Y. & Goujet, D. 1997: Nouveaux poissons acanthodiens du Dévonien du Spitsberg. *Geodiversitas 19*, 505–513.

Gagnier, P.-Y. & Wilson, M.V.H. 1995: New evidences on jaw bones and jaw articulations in acanthodians. *Geobios Mémoire Special 19*, 137–143.

Goodrich, E.S. 1909: Vertebrata craniata (First fascicle: cyclostomes and fishes). *In* Lankester, R. (ed.): *A Treatise on Zoology*. London.

Goodrich, E.S. 1930: *Studies on the Structure and Development of Vertebrates*. MacMillan, London.

Gross, W. 1947: Die Agnathen und Acanthodier des obersilurischen Beyrichienkalks. *Palaeontographica A 96*, 91–161.

Gross, W. 1957: Mundzähne und Hautzähne der Acanthodier und Arthrodiren. *Palaeontographica A 109*, 1–40.

Gross, W. 1967: Über das Gebiss der Acanthodier und Placodermen. *Journal of the Linnean Society (Zoology) 47*, 121–130.

Gross, W. 1971: Downtonische und dittonische Acanthodier-Reste des Ostseegebietes. *Palaeontographica A 136*, 1–82.

Hancock, A. & Atthey, T. 1868: Notes on the remains of some reptiles and fishes from the shales of the Northumberland coal field. *Annals and Magazine of Natural History 1*, 266–278, 346–378, pl. XIV–XVI.

Hanke, G.F., Davis, S.P. & Wilson, M.V.H. 2001a: New species of the acanthodian genus *Tetanopsyrus* from northern Canada, and comments on related taxa. *Journal of Vertebrate Paleontology 21*, 740–753.

Hanke, G.F. & Wilson, M.V.H. 2002: New teleostome fishes and acanthodian systematics. *Journal of Vertebrate Paleontology 22*, Suppl., 62A.

Hanke, G.F., Wilson, M.V.H. & Lindoe, L.A. 2001b: New species of Silurian acanthodians from the Mackenzie Mountains, Canada. *Canadian Journal of Earth Sciences 38*, 1517–1529.

Hermus, C.R. & Wilson, M.V.H. 2001: Early Devonian ischnacanthid acanthodians from the Northwest Territories of Canada. *Journal of Vertebrate Paleontology 21*, Suppl., xxA.

Hussakof, L. 1913: Description of four new Paleozoic fishes from North America. *Bulletin of the American Museum of Natural History 32*, 245–250.

Hussakof, L. & Bryant, W.L. 1918: Catalog of the fossil fishes in the Museum of the Buffalo Society of Natural Sciences. *Bulletin of the Buffalo Society of Natural Sciences 12*, 346 pp.

Huxley, T.H. 1861: Preliminary essay upon the systematic arrangement of the fishes of the Devonian epoch. *Memoirs of the Geological Survey of the United Kingdom Decade 10*, 1–40.

Janvier, P. 1977: Les poissons dévoniens de l'Iran central et de l'Afghanistan. *Mémoires de la Société Géologique du France 8*, 277–289.

Janvier, P. 1996: *Early Vertebrates*. Oxford University Press, Oxford.

Janvier, P. & Suarez-Riglos, M. 1986: The Silurian and Devonian vertebrates of Bolivia. *Bulletin de l'Institut Français d'Études Andines XV*, 73–114.

Jessen, H. 1973: Weitere Fischreste aus dem oberen Plattenkalk der Bergisch-Gladbach – Paffrather Mulde (Oberdevon, Rheinisches Schiefergebirge). *Palaeontographica A 143*, 159–187.

Karatajute-Talimaa, V. 1998: Determination methods for the exoskeletal remains of Early Vertebrates. *Mitteilungen aus dem Museum für Naturkunde in Berlin, Geowissenschaftliche Reihe 1*, 21–52.

Karatajute-Talimaa, V. & Smith, M.M. 2002: Early acanthodians from the Lower Silurian of Asia. *Transactions of the Royal Society of Edinburgh: Earth Sciences 93*, 277–299.

Lindley, I.D. 2000: Acanthodian fish remains from the Lower Devonian Cavan Bluff Limestone (Murrumbidgee Group), Taemas district, New South Wales. *Alcheringa 24*, 11–35.

Lindley, I.D. 2002: Lower Devonian ischnacanthid fish (Gnathostomata: Acanthodii) from the Taemas Limestone, Lake Burrinjuck, New South Wales. *Alcheringa 25*, 269–291.

Long, J.A. 1986: New ischnacanthid acanthodians from the Early Devonian of Australia, with a discussion of acanthodian interrelationships. *Journal of the Linnean Society (Zoology) 87*, 321–339.

Long, J.A., Burrow, C.J. & Ritchie, A. in press: A new ischnacanthid acanthodian from the Upper Devonian Hunter Formation near Grenfell, New South Wales. *Alcheringa 28*. In press.

Mallatt, J. 1984: Early vertebrate evolution: pharyngeal structure and the origin of gnathostomes. *Journal of Zoology, London 204*, 169–183.

Mark-Kurik, E. 1985: *Actinolepis spinosa* n. sp. (Arthrodira) from the Early Devonian of Latvia. *Journal of Vertebrate Paleontology 5*, 287–292.

Miles, R.S. 1965: Some features in the cranial morphology of acanthodians and the relationships of the Acanthodii. *Acta Zoologica 46*, 233–255.

Miles, R.S. 1966: The acanthodian fishes of the Devonian Plattenkalk of the Paffrath Trough in the Rhineland. *Arkiv für Zoologi 18*, 147–194.

Miles, R.S. 1973: Articulated acanthodian fishes from the Old Red Sandstone of England, with a review of the structure and evolution of the acanthodian shoulder-girdle. *Bulletin of the British Museum (Natural History), Geology 24*, 113–213.

Murchison, R.I. 1839: *The Silurian System*. J. Murray, London.

Ørvig, T. 1967: Some new acanthodian material from the Lower Devonian of Europe. *Journal of the Linnean Society (Zoology) 47*, 131–153.

Ørvig, T. 1973: Acanthodian dentition and its bearing on the relationships of the group. *Palaeontographica A 143*, 119–150.

Otto, M. 1999: New finds of vertebrates in the Middle Devonian Brandenberg Group (Sauerland, northwest Germany). *Paläontologische Zeitschrift 73*, 113–131.

Owen, R. 1846: *Lecture on the Comparative Anatomy and Physiology of the Vertebrate Animals, Delivered at the Royal College of Surgeons of England. I, Fishes*. Longman, Brown, Green & Longmans, London.

Priem, F. 1911: Sur des poissons et autres fossiles du Silurien supérieur du Portugal. *Communicacoes da Commissao do Servico Geologico de Portugal 8*, 1–11.

Schwartz, J.H. 1999: *Sudden Origins: Fossils, Genes, and the Emergence of Species*. John Wiley & Sons, New York.

Smith, M.M. & Coates, M.I. 1998: Evolutionary origins of the vertebrate dentition: phylogenetic patterns and developmental evolution. *European Journal of Oral Sciences 106* (Suppl. 1), 482–500.

Spjeldnaes, N. 1967: Acanthodians from the Siluro-Devonian of Ellesmere Island. *In* Oswald, D.H. (ed.): *International Symposium on the Devonian System, II*, 807–813. Alberta Society of Petroleum Geologists, Calgary.

Traquair, R.H. 1894: Notes on Palaeozoic fishes. No. 1. *Annals and Magazine of Natural History series 6*(14), 368–374.

Valiukevicius, J.J. 1992: First articulated *Poracanthodes* from the Lower Devonian of Severnaya Zemlya. *In* Mark-Kurik, E. (ed.): *Fossil Fishes as Living Animals*, 193–214. Academy of Sciences of Estonia, Tallinn.

Valiukevicius, J.J. 1997: An unusually squamated acanthodian from Severnaya Zemlya. *In* Wilson, M.V.H. (ed): *Circum-Arctic Palaeozoic Vertebrates: Biological and Geological Significance, Buckow, 4–6 April 1997*, 25. Ichtyolrth Issues Special Publication 2, Edmonton.

Valiukevicius, J.J. 2003: Devonian acanthodians from Severnaya Zemlya Archipelago. *Geodiversitas 25*, 131–204.

Vergoossen, J.M.J. 1993: Jawbones in northern Siluro-Devonian sedimentary erratics: an appeal to collectors to register their finds. *Grondboor en Hamer 47*, 108–112 (in Dutch).

Wang, N.-Z. 1984: Thelodont, acanthodian, and chondrichthyan fossils from the Lower Devonian of southwest China. *Proceedings of the Linnean Society of New South Wales 107*, 419–441.

Watson, D.M.S. 1937: The acanthodian fishes. *Philosophical Transactions of the Royal Society (B) 228*, 49–146.

Wellburn, E.D. 1902: On the fish fauna of the Pendleside limestones. *Proceedings of the Yorkshire Geologic and Polytechnic Society 14*, 465–473.

White, E.I. 1961: The Old Red Sandstone of Brown Clee Hill and the adjacent area, II. Stratigraphy. *Bulletin of the British Museum (Natural History), Geology 5*, 243–310.

Woodward, A.S. 1891: *Catalogue of the Fossil Fishes in the British Museum (Natural History). Part II*. British Museum (Natural History), London.

Diplacanthid acanthodians from the Aztec Siltstone (late Middle Devonian) of southern Victoria Land, Antarctica

GAVIN C. YOUNG & CAROLE J. BURROW

Young, G.C. & Burrow, C.J. **2004 06 01**: Diplacanthid acanthodians from the Aztec Siltstone (late Middle Devonian) of southern Victoria Land, Antarctica. *Fossils and Strata*, No. 50, pp. 23–43. Australia. ISSN 0300-9491.

One articulated and several partial, semi-articulated specimens of acanthodians were collected in 1970 from the freshwater deposits of the Aztec Siltstone (Middle Devonian; Givetian), Portal Mountain, southern Victoria Land, Antarctica, during a Victoria University of Wellington Antarctic Expedition. The Portal Mountain fish fauna, preserved in a finely laminated, non-calcareous siltstone, includes acanthodians, palaeoniscoids, and bothriolepid placoderms. The articulated acanthodian specimens are the most complete fossil fish remains documented so far from the Aztec assemblage, which is the most diverse fossil vertebrate fauna known from Antarctica. They are described as a new taxon, *Milesacanthus antarctica* gen. et sp. nov., which is assigned to the family Diplacanthidae. Its fin spines show some similarities to spine fragments named *Byssacanthoides debenhami* from glacial moraine at Granite Harbour, Antarctica, and much larger spines named *Antarctonchus glacialis* from outcrops of the Aztec Siltstone in the Boomerang Range, southern Victoria Land. Both of these are reviewed, and retained as form taxa for isolated spines. Various isolated remains of fin spines and scales are described from Portal Mountain and Mount Crean (Lashly Range), and referred to *Milesacanthus antarctica* gen. et sp. nov. The histology of spines and scales is documented for the first time, and compared with acanthodian material from the Devonian of Australia and Europe. Distinctive fin spines from Mount Crean are provisionally assigned to *Culmacanthus antarctica* Young, 1989b. Several features on the most complete of the new fish specimens – in particular, the apparent lack of an enlarged cheek plate – suggest a revision of the diagnosis for the Diplacanthidae.

Key words: Acanthodii; Antarctica; Aztec Siltstone; Devonian; Givetian; new genus *Milesacanthus*; Diplacanthidae.

Gavin C. Young [gyoung@ems.anu.edu.au], Department of Earth and Marine Sciences, Australian National University, Canberra, ACT 0200, Australia

Carole J. Burrow [C.Burrow@uq.edu.au], Department of Zoology and Entomology, University of Queensland, QLD 4072, Australia

Introduction

Devonian fish were discovered in Antarctica during the British Antarctic "Terra Nova" expedition of 1910–1913, when T. Griffith Taylor's party explored the coast of Victoria Land in the summer of 1910–1911. These were the first Devonian fossils, and the first fossil vertebrates, to be discovered on the Antarctic continent. Fossil fish remains discovered in glacial moraine at Mount Suess, near the mouth of the Mackay Glacier in Granite Harbour

---

Abbreviations used in figures

adfs, anterior dorsal fin spine; adfw, web of anterior dorsal fin; admfs, admedian fin spine; anfs, anal fin spine; apectfs, allochthonous pectoral fin spine; br, branchiostegal rays; c, canal; cb, thin canals of base; cc, central spine cavity; cf, caudal fin; cnb, circumnaral bones; cob, circumorbital bones; dc, dorsal "cone" of procoracoid; dp, indeterminate dermal plate (?sarcopterygian scale); fs, fin spine; gr, crown groove; gz, crown growth zones; il, inner layer of spine; ins, inserted part of spine; ioc.pt, post-orbital part, infra-orbital sensory line; lms, left mandibular splint; lc, main lateral sensory line; lsc, left scapulocoracoid; ol, outer layer of fin spine; os, scales from another fish; osc, scapulocoracoid from another fish; p, pores; pc, canals of pore canal system; pdfs, posterior dorsal fin spine; pdfw, web of posterior dorsal fin; pectfs, pectoral fin spine; pectfw, web of pectoral fin; pelvfs, pelvic fin spine; pfc, profundus canal; ppfs, pre-pelvic fin spine; pro, procoracoid; r, crown ridge; rms, right mandibular splint; rsc, right scapulocoracoid; scc, subcostal canal; sl, sensory line; sp, indeterminate spine; tess, head tesserae; vp, ventral plate of procoracoid.

(locality 2 in Young 1988, fig. 3), were assumed by Debenham (1921) to derive from the thick sequence of sedimentary rocks called the "Beacon Sandstone", well exposed in the region of the lower Ferrar and Taylor Glaciers. Woodward (1921) described the material, identifying eight taxa of Devonian fishes.

The first *in situ* material was collected in the Skelton Névé region, over 100 km to the south, during the Trans-Antarctic Expedition of 1955–1958 (Gunn & Warren 1962). This small collection was described by White (1968). Many new fossil localities were discovered in the same area during the 1968–1969 summer field season of the New Zealand Antarctic Research Program (NZARP). The main fossil fish collection, from the Aztec Siltstone, Taylor Group, Beacon Supergroup (see McKelvey *et al.* 1972; McPherson 1978), was made by a Victoria University of Wellington Antarctic Expedition (VUWAE 15) in the 1970–1971 field season. Later expeditions (1976–1977, 1988–1989, 1991–1992) also collected material from new localities in the Cook Mountains (Woolfe *et al.* 1990; Long & Young 1995, fig. 1), 100 km to the south of previously known sites (M. A. Bradshaw, NZARP, event 33; J. A. Long, NZARP–ANARE). Descriptions, based mainly on material in the 1970–1971 VUWAE 15 collection, show the Aztec fish fauna to be one of the most diverse known assemblages of Middle–Late Devonian age, including representatives of most of the major vertebrate groups: thelodont agnathans (Turner & Young 1992), chondrichthyans (Young 1982; Long & Young 1995), placoderms (Ritchie 1975; Young 1988; Long 1995), sarcopterygians (Young *et al.* 1992), and acanthodians, actinopterygians and dipnoans (Campbell & Barwick 1987: fig. 2; Young 1989a, b, 1991). The last three groups are not fully described. Forty-two taxa were listed in the Aztec fauna by Long & Young (1995, table 1).

In this paper we describe the only articulated acanthodian remains from the Aztec fish fauna, plus other isolated material, collected during the 1970–1971 expedition, and we revise earlier descriptions of acanthodian remains by Woodward (1921) and White (1968). Woodward (1921) compared isolated acanthodian scales from the Granite Harbour material with those of *Cheiracanthus murchisoni* from the Middle Devonian of Scotland. White (1968) described acanthodian spines of several types from two localities (Boomerang Range, Mount Crean), one of which ("*Cosmacanthus*? sp.") is probably a placoderm spinal plate (Denison 1978). Ribbed spines from the upper fossiliferous horizon in the Boomerang Range were referred to a new species of the genus *Gyracanthides*, originally described by Woodward (1906) from the Early Carboniferous of Mansfield in Victoria, Australia. This taxon was recently redescribed by Warren *et al.* (2000). White (1968) also erected a new genus and species,

Antarctonchus glacialis White, and correctly identified as acanthodian spines the fragments originally described by Woodward (1921) as an antiarch (*Byssacanthoides debenhami* Woodward). Other acanthodians in the Aztec fish fauna include *Culmacanthus antarctica* Young, 1989b from Mount Crean, and a set of articulated jaws (Long & Young 1995, table 1, "Ischnacanthid gen. indet."). Elsewhere in Antarctica, acanthodian spines (*Machaeracanthus*) have been recorded from the Early Devonian of the Ohio Range (Young 1986) and the Ellsworth Mountains (Young 1992).

Localities, stratigraphic occurrence and age

A locality map for the 24 fossil fish localities of the 1970–1971 and earlier field seasons in the Granite Harbour–Skelton Névé region of southern Victoria Land was published by Young (1988, fig. 3), when full locality details were given. The same map has been published by Young (1989a, fig. 2; 1991, fig. 15.4), Young *et al.* (1992, fig. 2), and Turner & Young (1992, fig. 1). The material dealt with here comes from localities 2, 4, 8, 11, 12 and 20 on that map.

Young (1989a, 1991) noted that remains of ribbed acanthodian spines occurred at most Aztec fish localities, but this is not accurate, and a more detailed survey of the whole collection indicates that they are confined to the lower and middle horizons of the Aztec sequence. The lowest record is an incomplete spine (Young 1991, fig. 15.7a) from the top of the Beacon Heights Orthoquartzite at Mount Fleming (locality 4 in Young 1988, fig. 3). *Antarctonchus glacialis* White, 1968 was based on fin spine impressions from two localities. The type locality (MS2 of Gunn & Warren 1962) is in the Boomerang Range (Young 1988, fig. 3, locality 20), from a horizon probably some 70 m above the base of the formation (Young 1988, p. 9). The only spine from the second locality (MS5 of Gunn & Warren 1962; Mount Crean, Lashly Range) may not belong to this species. The upper horizon of the original Mount Crean collection may approximate to Unit 8 of Section L2 of Askin *et al.* (1971), but correlations are uncertain, as discussed by Young (1988, pp. 12, 13). Among the extensive fish material collected in 1970–1971 Young (1988, pp. 12, 13) noted indeterminate acanthodian remains from collection sites MC1–3 at the Mount Crean locality, and Units 4 and 14 at Portal Mountain (localities 8, 11, 12 in Young 1988, fig. 3). This is the material described in this paper.

The Aztec fish fauna is now considered to be somewhat older than the initial age assessment of Upper Devonian (see discussion in Young 1993). It was referred to the late Middle Devonian (Givetian) by Young (1996), and

Turner (1997) considered the thelodont *Turinia antarctica* to be of early Givetian age; this taxon identifies the lowest two zones in the biostratigraphic scheme for the Aztec sequence proposed by Young (1988, fig. 5). The material referred below to *Milesacanthus antarctica* gen. et sp. nov. comes from equivalent horizons at Mount Crean, and also from the slightly higher "*portalensis*" biozone at Portal Mountain. The type locality for *Antarctonchus glacialis* White is higher in the Aztec sequence, at a level which has yielded spines of the acanthodian *Gyracanthides warreni* White, 1968, and phyllolepid placoderm remains in the Boomerang Range. These occurrences (70 m above the base of the Aztec Siltstone) were considered to be anomalously low by Young (1988, p. 14), because both phyllolepids and *Gyracanthides* are indicators of the youngest Aztec biozone. However Long & Young (1995) recorded both rare phyllolepid remains and *Gyracanthides* from a similar level in the Aztec sequence at the "Fish Hotel" site in the Cook Mountains.

On present evidence, all of the acanthodian material dealt with here is probably Givetian in age, with a possible stratigraphic separation between lower (*Byssacanthoides*, *Milesacanthus*) and upper (*Antarctonchus*, *Gyracanthides*) acanthodian occurrences in the Aztec sequence. Smooth ribbed spines with radiating internal structure similar to those described below occur in the European Devonian, but on present data do not improve the age resolution for the Aztec fish fauna, as they are both older (*Onchus overathensis*, Pragian) and younger (*Devononchus concinnus*, early Frasnian) than the Givetian age assumed here.

Material and methods

Most of the fish material from the Aztec Siltstone is preserved as light-coloured bone in a darker siltstone or fine sandstone matrix, and has been prepared by mechanical removal of the matrix or by removal of bone to give impressions for latex rubber casting. An articulated acanthodian on specimen ANU V773 was mechanically prepared some years ago by Dr A. Ritchie (Australian Museum), and we are uncertain about some important structures that could have been lost during preparation. Thin sections often show well-preserved histology in the fin spines, and a few calcareous beds dissolved in acetic acid have yielded scales which are well preserved both morphologically and histologically. In this paper we deal only with the 1970–1971 and earlier collections of acanthodian material, plus some acid-prepared scales from the Mount Crean locality collected later, and made available by Dr John Long (Western Australian Museum).

The material described or mentioned here is housed in the Department of Earth and Marine Sciences, Australian National University, Canberra (ANU V), the Australian Museum, Sydney (AMF), the National Museum of Victoria, Melbourne (NMV P), the Western Australian Museum, Perth (WAM), the Natural History Museum, London (NHM P), and the Institute of Geological and Nuclear Sciences, Wellington, New Zealand (ex-New Zealand Geological Survey; GS).

Systematic palaeontology

Class Acanthodii Owen, 1846

Family Diplacanthidae Woodward, 1891

Revised diagnosis. – Acanthodians with: a short mouth and cheek region; mandibular splint (*sensu* Watson 1937); no teeth or dermally ossified tooth plates; a high cylindrical scapular shaft and triangular posterior flange on the scapulocoracoid; some with a procoracoid; circumorbital bones plus a pre-opercular, or cheek plate and/or up to five pairs of flattened branchiostegal rays; ornamented scales with acellular dentine and wide canals in the crowns, and acellular bases; fin spines ornamented with smooth longitudinal ribs paralleling the leading edge of the spine; dermal shoulder girdle sometimes incorporates paired pinnal plates; admedian fin spines ventromedial to the pectoral fin spines; some with one pair of pre-pelvic series fin spines (*sensu* Wilson 1998); unpaired and pelvic fin spines deeply inserted into the body musculature.

Remarks. – Three orders are "classically" recognised within the Acanthodii: the Climatiiformes, Acanthodiformes and Ischnacanthiformes Berg, 1940 (Climatiida, Acanthodida, Ischnacanthida of some other workers; e.g. Denison 1979). Of these, the Climatiiformes may be paraphyletic (Janvier 1996), while one dorsal fin spine and dentigerous jaw bones of dermal origin may be synapomorphies of the Acanthodiformes and Ischnacanthiformes, respectively. Hanke *et al.* (2001) revived the order Diplacanthiformes Berg, 1940, first established to include only the Diplacanthidae Woodward, 1891, with diagnostic characters (as listed by Berg 1940) based on *Diplacanthus* Agassiz, 1844. Hanke *et al.* (2001) added the families Gladiobranchidae and Tetanopsyridae (Bernacsek & Dineley 1977; Gagnier *et al.* 1999), but did not revise the diagnosis for the Diplacanthiformes.

Shared derived characters of the three "diplacanthiform" families proposed by Hanke *et al.* (2001) are: toothless blade-like jaws, some enlarged circumorbital plates, scapulocoracoid with enlarged posterior flange, and dermal hyoidean gill covers. Hanke *et al.* (2001)

heralded a forthcoming cladistic analysis of the acanthodians, which might clarify the relationships between the Diplacanthidae, Tetanopsyridae, Gladiobranchidae and Culmacanthidae Long, 1983. Some characters listed by Hanke *et al.* (2001) to unite the first three families may be tenuous, with many climatiids also having posterior flanges on the scapulocoracoid, branchiostegal rays (equivalent to dermal hyoidean gill covers), and circumorbital plates of a similar size to those of *Tetanopsyrus*. Hanke *et al.* (2001) differentiated the Tetanopsyridae from the Diplacanthidae by the former lacking pre-pelvic and admedian spines, dermal plates in the shoulder girdle, cheek plates, and mandibular splints, and in having ossified jaws (the "toothless plates" in the Tetanopsyridae may be part of the perichondral ossification of the jaw cartilages, thus differing from the dermal dentigerous jaw bones of ischnacanthiforms). The Gladiobranchidae differ from the Diplacanthidae in having pre-pectoral spines, and lacking scapular shafts with a circular cross-section. The Culmacanthidae (only genus, *Culmacanthus*) is distinguished from the Diplacanthidae by the very large cheek plate, three distinctively ornamented dermal shoulder girdle bones, and scale morphology. *Culmacanthus* also lacks pre-pectoral, pre-pelvic and admedian (= "first intermediate") spines. The "admedian" spine is positioned ventrally between the pectoral spines, and in *Diplacanthus striatus* is attached to a pinnal plate. Miles (1973a) referred to this spine as the "first intermediate" spine, but it seems unlikely that it was part of the pre-pelvic series (*sensu* Wilson 1998; new terminology replacing "intermediate" fin spines).

At this time, little concrete support exists for uniting these families in a higher group Diplacanthiformes. Long (1983) united the Culmacanthidae and Diplacanthidae in the suborder Diplacanthoidei, first erected by Miles (1966) for the single family Diplacanthidae, but Hanke *et al.* (2001, p. 752) considered many features of *Culmacanthus* to be "completely different from those of diplacanthids" (see below). As we consider that the higher level classification of acanthodians outside the Ischnacanthiformes and Acanthodiformes is currently unresolved, we have not assigned the Diplacanthidae to an order. Our amended diagnosis is modified from Denison (1979) to incorporate new features described below. We include three genera (represented by articulated remains) in the Diplacanthidae: *Diplacanthus*, *Rhadinacanthus*, and *Milesacanthus* gen. nov.

Genus *Milesacanthus* nov.

Type species. – *Milesacanthus antarctica* sp. nov.

Etymology. – After Dr Roger Miles, who has made a major contribution to the study of Palaeozoic fishes generally,

and to acanthodian fishes in particular; and "*akantha*" (Greek) for "thorn".

Diagnosis. – Diplacanthid acanthodians with: four pairs of flattened, ornamented branchiostegal rays; elongated bones edging the orbit posteriorly; scapulocoracoid expands ventrally to a relatively short triangular base; procoracoid comprising an upper cone and ventral plate. Paired pectoral, admedian, and pelvic fin spines; fin spines up to 8 cm long with a subcircular/subtriangular cross-section, and up to six smooth longitudinal ribs per side decreasing in width posteriorly, plus a wider leading edge rib. Unpaired and pelvic spine fin webs extend to the tip of the spine; anterior dorsal and anal fin spines are slightly curved and of comparable length; pelvic and posterior dorsal fin spines are slightly shorter and relatively straight with four or five ribs per side; admedian spines are short and conical. All spines have a large central pulp cavity, lack a subcostal canal, with narrow canals radiating out from the central cavity. Flank scales are ornamented with 14–24 subparallel ridges or grooves extending from the anterior edge to at least midcrown; scales have a pore canal system in the anterior part of the crown with pore openings in the crown grooves and the lower neck, fine dentine tubules and canals rising up from the base/crown boundary and curving over into the upper horizontal parts of the crown growth zones.

Milesacanthus antarctica sp. nov.
Figs. 1–5, 6H, O–R

Synonymy. –

1921   *Cheiracanthus* sp. – Woodward, p. 56, pl. 1, figs. 12, 13
1968   *Antarctonchus glacialis* – White, pp. 11, 12 *in pars*
1988   "acanthodians" – Young, pp. 12, 13
1989a   "partly articulated material ..., with fin webs, spines, and much of the scale cover preserved, resembles the Middle Devonian Scottish form *Diplacanthus*" – Young, p. 51
1989a   "acanthodian scales from Mount Fleming" – Young, fig. 4C
1991   "acanthodian ..." with "diplacanthid affinity" – Young, p. 549
1992   "some additional acanthodians" – Turner & Young, p. 90
1993   "*Cheiracanthoides comptus*" – Young *et al.*, p. 248
2002   "diplacanthid acanthodian" – Burrow & Young, p. 194

Etymology. – From Antarctica.

Holotype. – "Fish 1" on Portal Mountain sample ANU V773.

Other material. – New material referred to this species comes from two localities and horizons at Portal Mountain, and several collecting sites at Mount Crean (specimen numbers listed in the next section). Isolated elements include fin spines (ANU V775, 776, 778, 779, 860, 880, 890, 968, 969, 1178, 2165, 2174, AMF 55623, 55859, 55886, 55873; Figs. 4B–D, 5C), scale patches (Fig. 4D) and individual scales (ANU V893; Fig. 6Q, R), thin sections of scales and spines (WAM 03.1.1, 2, ANU V2163; Figs. 3C, D, 5A, B, D, 6H, O, P), and a possible neurocranial element (ANU V774; Fig. 4E).

Localities and horizon. – Material from Portal Mountain (locality 12 of Young 1988, fig. 3; latitude 78°7.2'S, longitude 159°24'E), including articulated fish ANU V773, and various isolated elements (ANU V774–6, 778, 779, AMF 55623) came from Unit 14, Section P1 of Askin *et al.* (1971). Isolated elements (ANU V860, 1178, 2163, 2165, 2174) from a higher horizon (locality 11 of Young 1988, fig. 3; latitude 78°7.2'S, longitude 159°23.5'E) are referred to Unit 17, Section 10 of Barrett & Webb (1973).

Isolated acanthodian remains from Mount Crean, Lashly Range (locality 8 of Young 1988, fig. 3; latitude 77°53'S, longitude 159°33'E) came from lower units in the Aztec Siltstone (Section L2 of Askin *et al.* 1971) at collecting sites indicated by Young (1988, fig. 4) as follows: MC1 (ANU V880), MC2 (ANU V890, 893), MC3 (ANU V968, 969, WAM 03.1.1, 03.1.2). AMF 55859, 55873, 55886 are also referred to these localities (precise position unspecified).

Diagnosis. – As for genus (only species).

Remarks. – Two taxa in the Aztec fauna have been based on similar fin spines: *Byssacanthoides debenhami* Woodward and *Antarctonchus glacialis* White. Spine morphology on the new articulated specimens supports some of the criteria used by White (1968) to distinguish these taxa (see below), and the new fin spines seem more similar to *Byssacanthoides* than to *Antarctonchus*. However, the provenance of the type material of *Byssacanthoides debenhami* Woodward is unknown, and they are very small isolated fragments displaying few characters, but showing some differences (wider leading edge rib, more rounded cross-section) to spines in our new material. The spines named *Antarctonchus* are larger, and come from a higher horizon in the Aztec sequence. Given the vast amount of information provided by articulated specimens compared with isolated spines or spine fragments, it is appropriate to restrict *Byssacanthoides debenhami* and *Antarctonchus glacialis* for use as form taxa for isolated spines. In addition, the absence of a large cheek plate in the new articulated specimens shows that they do not belong to *Culmacanthus antarctica* Young, 1989b, even though *Culmacanthus* also had similar fin spines with

smooth longitudinal ribs (Long 1983). On this basis we assign the new articulated specimens, and most of the new isolated material, to a new taxon *Milesacanthus antarctica* gen. et sp. nov. Apart from *Gyracanthides warreni* White, 1968, the available evidence suggests that the other named acanthodian taxa in the Aztec fish fauna may be closely related.

Description. – The most important acanthodian specimen in the 1970–1971 collection is ANU V773 from Portal Mountain (Fig. 1), an accumulation of partly articulated acanthodian remains on several bedding planes in a finely laminated and fissile greenish-grey siltstone. It comprises four main pieces (A–D), together with numerous small flakes and broken scale patches. Pieces A and C (21 and 17 cm across, respectively) are less than 7 mm thick, and fit together across a straight rock fracture to form the part of the specimen. Acanthodian remains pass off the edge of piece C (Fig. 1A). Pieces B and D fit together across the same straight fracture to form the counterpart (Fig. 1B). The latter are much thicker (up to 23 mm) and larger (21, 24 cm across), with dense patches of squamation and disoriented spines extending outside the area covered by pieces A and C. At least 11 fin spines occur on pieces C and D, of which only one clearly belongs to the best preserved individual (the holotype: "Fish 1", Fig. 1A). This specimen has the anterior part of the body well displayed on piece A, and forms the basis of the description below.

Most of the right side of the holotype (Fish 1) is preserved on pieces A and C. Some of its fin spines are disrupted, and remains of at least three other individuals lie at angles across the tail region (Fig. 1A). Scattered scales and spines occur outside this accumulation, and on different bedding planes on the specimen. Piece A was glued together across a fracture passing up through the anterior dorsal fin spine. Behind the fracture, the split along the bedding plane passed within the body of Fish 1, more or less separating the left and right sides of the squamation, which is mainly preserved as broken sections through the middle of the scales. A few areas show ridged impressions of the scale crowns. Scale preservation is similar in front of the fracture, again mostly showing broken sections through the middle of the scales. This anterior part was originally enclosed in the rock matrix, and was prepared out mechanically (by A. Ritchie, Australian Museum) after backing with resin. The preserved part comprises the right side and some of the left side of the fish compressed together. In this circumstance it is unclear whether broken scales are from the left or right side, although some patches preserve the scale crowns facing upwards, so must be from the left. The counterpart (piece B; Fig. 1B) preserves the left side behind the level of the anterior dorsal fin spine, but with missing scale patches, and two displaced fin spines assumed to come

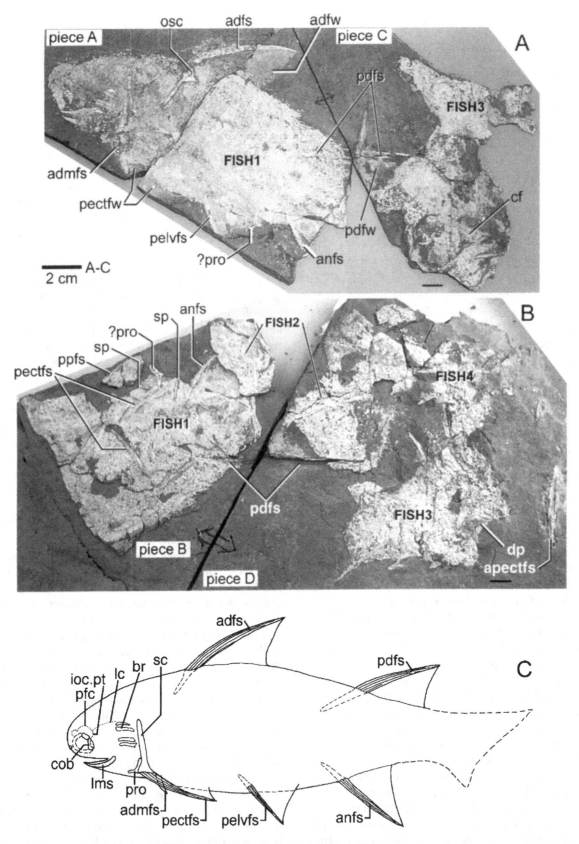

Fig. 1. Milesacanthus antarctica gen. et sp. nov. A, B: Sample ANU V773 (Portal Mountain), showing at least four partially articulated acanthodian fishes, on four pieces (A–D) of siltstone slab making up the part (A) and counterpart (B) of the sample. C: Outline reconstruction, based on the holotype (Fish 1 in A, B).

from the pectoral girdle. Fishes 2–4 (Fig. 1A, B) are represented by fragmentary and/or disarticulated remains on the part and counterpart (pieces B, C, D).

The holotype is laterally compressed, and its total length is estimated at 220 mm. The maximum preserved depth (just behind the anterior dorsal fin spine) is 69 mm. Much of the caudal region is missing or obscured behind the insertions for anal and posterior dorsal fin spines (anfs, pdfs, Fig. 1A). However, the free end of the posterior dorsal fin spine and the outline and some squamation from the caudal tip are traceable on the adjoining piece C (cf, Fig. 1A).

Head and branchial region

The elements of the head preserved *in situ* on piece A include parts of the squamation from both sides (mainly the right), both mandibular splints, three or four of the post-orbital bones, and four branchiostegal rays (Fig. 2A). The rounded rostrum is covered with irregular polygonal tesserae (Fig. 2B) which range from 0.5 to 1.0 mm wide, and extend about 30 mm back along the dorsal midline, to the branchiostegal rays laterally, and just posterior to the mandibular splints along the ventral midline. At least one semicircular nasal bone (cnb, Fig. 2B) is discernible among the anterior-most tesserae. A patch of scales, preserved between the nasal region and the mandibular splints and in front of the circumorbital bones, is apparently continuous with the rostral margin, and forms a posteroventrally projecting lobe of scales. Scale crowns are aligned along the margin, with the anterior of each scale facing posteroventrally. They are shaped like flank scales, so this may be a displaced patch, perhaps from another fish (?os, Fig. 2B). The posterior left, and possibly some of the right circumorbital bones, are preserved, one bearing part of a sensory line. The orientation and rectilinear shape of some of the tesserae suggests that they bordered some of the sensory lines of the head (sl, Fig. 2B). The profundus canal (pfc, Fig. 1C), post-orbital part of the infra-orbital sensory canal and pre-scapular section of the main lateral sensory canal can be identified (terminology of Denison 1979, fig. 4). The whole of the right mandibular splint and the anterior half plus ventral edge of the left splint are preserved (rms, lms, Fig. 2A).

There are no big patches of missing scales in the cheek region, and no evidence of a large cheek plate such as occurs in *Culmacanthus*. Four smaller elements positioned posterodorsal to the mandibular splints are interpreted as branchiostegal rays (br, Fig. 2A, C). The lower two rays are very closely associated, and apparently in contact at their posterior ends, and about half way along their length, with elsewhere a single scale row between them. The lower ray is 9 mm long but possibly incomplete anteriorly, and the one above is 10 mm long and 1.5 mm wide at its posterior end. The upper two rays

are separated by a gap of about 3 mm, which contains many scales, so it seems unlikely that there could have been an intervening element that has been lost. The upper rays are shorter (uppermost 5.5 mm long, but possibly incomplete; one beneath 8.5 mm long). They appear to have been in contact anteriorly, behind which they are separated by one, two, and then several rows of scales. Behind this the scales are disrupted. Beneath the lower margin of the upper pair of rays is a closely associated row of slightly enlarged scales, apparently *in situ*. The exposed surfaces of all four rays are smooth, but the anterior edges of the upper pair where the bone tissue is mainly missing shows irregular pyritic filling suggesting that the opposite surface was ornamented. The absence of ornament on the exposed surfaces, and the apparent *in situ* arrangement of surrounding scales supports the interpretation that they all come from the right side, and that all of the left branchiostegal rays were lost during preparation.

Pectoral region

Normal flank scales are found posterior to the branchial region; they are largest (about 1 mm long) just behind the uppermost pair of rays, decreasing in size dorsally to about 0.5 mm at the midline. The scapulocoracoids are about 5 mm behind the branchiostegals, with the left incompletely preserved slightly in front of the right (lsc, rsc, Fig. 2A). The narrow scapular shaft is about 20 mm high, with a circular cross-section expanding slightly towards the base to form a posteriad triangular flange. A bare patch behind the left scapulocoracoid might indicate where more of the posterior flange was lost during preparation. Probably it resembled more the scapulocoracoid of *Culmacanthus*, rather than that of *Diplacanthus* with its large posterior flange (Miles 1973a, fig. 40; cf. Long 1983, fig. 6).

The squamation is disrupted or missing behind the mouth region, probably caused by detachment and displacement of the pectoral fin spines connected to the endoskeletal girdle as the carcass settled and was buried in the sediment. However, the proximal part of the pectoral fin web is preserved (pectfw, Fig. 1A), with its leading edge meeting the flank about 55 mm from the tip of the rostrum. The pectoral fin spines are not preserved *in situ*. The leading edge of the pectoral fin web is undisrupted despite the absence of the fin spine, but the pectoral fin spine is not attached to the web in some other acanthodians (e.g. *Brochoadmones* Gagnier & Wilson, 1996), which might explain this. Two other spines preserved on and under the body squamation on the counterpart (piece B) are interpreted as pectoral fin spines (pectfs, Fig. 1B). Ventral to the scapulocoracoids is a short conical ribbed spinelet, 2 mm in diameter at the base and 7 mm long (admfs, Figs. 1A, 2A), with a single row of *in situ* scales running outside its ventral margin at a deeper level in the matrix. Thus, this spine appears to overlie the

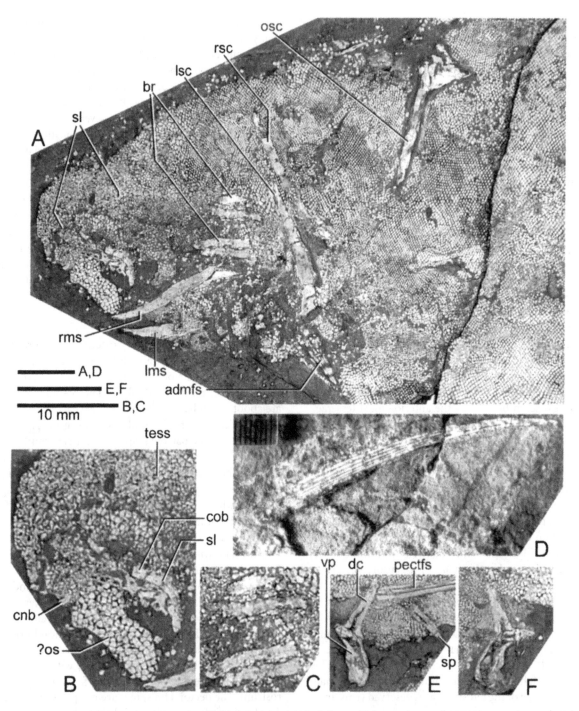

Fig. 2.   Milesacanthus antarctica gen. et sp. nov. A: Detail of the anterior end of the holotype (Fish 1 on piece A of ANU V773; see Fig. 1A). B: Enlargement of the orbital region shown in A. C: Enlargement of the dermal branchiostegal rays ("br" in A), internal sides exposed. D: Anterior dorsal fin spine and fin web of Fish 1 on piece A ("adfs", "adfw", Fig. 1A). E, F: Displaced procoracoid in part and counterpart on pieces A and B of ANU V773 ("?pro", Fig. 1A, B).

squamation, and is interpreted as a right admedian spine (admfs, Fig. 1C).

Several elements thought to derive from another fish are preserved along the dorsal surface of the head, posterior to the scapulocoracoids. These include a patch of scales with crowns pointing in the opposite direction

to the surrounding scales, a small thin fin spine fragment, a small plate, and a complete scapulocoracoid (osc, Fig. 2A). The latter is oriented upside down just below the anterior dorsal fin, with its lateral face and the ventral medial edge exposed. The triangular area at its base is the posterior flange.

Pelvic and caudal regions

A distance of 10 mm behind the posterior limit of the pectoral fin web is another fin spine and extensive fin web (Fig. 1A). The corresponding spine and fin web of the left side are less completely preserved on piece B (Fig. 1B). As there is no other fin or spine between the anal fin spine and the pectoral fin web, we assume that this is the pelvic fin and spine.

Adjacent and posterior to this fin on piece B are two small spine fragments, presumably displaced, which lie on either side of another displaced structure which is difficult to interpret (?pro, Fig. 1A). This complex has been split through the centre in separating the part and counterpart, revealing its inner structure (Figs. 1A, B, 2E, F). It is tripartite, comprising an upper thick-walled "cone", 7 mm high with a 2 mm basal diameter, a short hollow central part about 3 mm long, which points back at right angles to the cone and abuts its posterior wall, and a hollow, subrectangular, ventral plate, making an angle of about 120° with the central part. The latter appears to be the base of a fin spine, from which the free part of the spine has broken off. The cone and plate appear to be perichondrally ossified.

As this element is in the pelvic region, one possibility is that it represents the pelvic girdle, either slightly displaced to a position midway between the pelvic and anal spines, or perhaps from another fish. Under this interpretation, the upper cone could be the pelvic equivalent of the scapular part of the pectoral girdle, and the ventral part might correspond to the "pelvic basal plate". Acanthodian pelvic girdles have been described only rarely. Acanthodidids *Acanthodes gracilis* and *A. bridgei* have a spoon-shaped, perichondrally ossified, pelvic basal plate (Zidek 1976, fig. 8A, B). At least one specimen of *Ischnacanthus gracilis* is purported to have a cartilaginous pelvic girdle of indeterminate shape (Watson 1937). Dean (1907, fig. 19) labelled elements as pelvic girdles on *Diplacanthus striatus*, but these were actually scapulocoracoids. Among other early gnathostomes, the acanthothoracid placoderm *Murrindalaspis* had a tripartite structure combining fused lateral, spinal, and ventral dermal plates attached to an endoskeletal girdle (Long & Young 1988, figs. 6–8), an example of a serial homologue of the internal structuring of its pectoral girdle.

Against the pelvic girdle interpretation is the undisrupted preservation of the fin spine and web on piece A, which we interpret as the pelvic fin. Alternatively this element could be a displaced procoracoid complex from the pectoral girdle, as there are presumed displaced pectoral spines nearby, and there are no pectoral spines preserved *in situ*. The procoracoid has a similar shape, with dorsal cone-shaped and ventral plate components, in *Diplacanthus striatus* (Denison 1979, fig. 20). There is no evidence of an attached dermal plate comparable with those of *Diplacanthus striatus* and *Gyracanthides murrayi*

(Warren *et al.* 2000, fig. 5A, B). *Diplacanthus horridus* Woodward, 1892 and *D. ellsi* Gagnier, 1996 also lack dermal plates. The spine base could belong to either a pectoral or an admedian fin spine. Yet another possibility is that the cone is the left admedian fin spine, and the plate is the pinnal plate separating the pectoral and admedian fin spines as in *Diplacanthus striatus*, but this seems less likely as there is no indication of ornament on the cone or plate, and the cone is contiguous with the fin spine base fragment rather than separated from it by the plate. Considering all three possibilities, interpretation as a procoracoid complex seems most likely.

The distal part of the anal fin spine (anfs, Fig. 1A, B) and its intact fin web, are obscured by a patch of scales from Fish 2 on the counterpart (piece B). Behind this, the caudal region is minimally preserved on piece C, where it is partly obscured by Fish 2, with part of the caudal fin web projecting posteriorly (cf, Fig. 1A).

Squamation and sensory lines

The squamation of this fish comprises about 450 diagonal rows of scales, with about 150 scales/row on each side. As on most acanthodians, the head is covered with tesserae showing areal growth zones, not the normal "box-in-box"-type acanthodian scales. Unfortunately, as the tesserae are all split horizontally, the surface ornament is not visible. The internal structure of the head tesserae, as revealed by natural horizontal sections (Fig. 3A), shows wide vascular canals radiating through the posterior two-thirds of the crown, with dentine tubules extending out from them.

Scales bordering the sensory lines on the head appear slightly larger and more rectangular than the surrounding tesserae (Fig. 2A, B). Only parts of the sensory lines are recognised: the post-orbital part of the infra-orbital canal, the profundus canal extending over the orbit, and the main lateral canal running back from the dorsal end of the profundus canal and above the dorsal edge of the uppermost branchiostegal ray (Fig. 1C). It is not clear, with only inner surfaces preserved, whether these scales could have partly enclosed the sensory canals, as in *Diplacanthus* and many acanthodids (Denison 1979, p. 11). There are some enlarged tesserae or small plates in the nasal region.

Posterior to the fracture line which separates the head/pectoral region from the body, most of the scales on both the part and the counterpart have their inner surfaces exposed, or have been cracked through the middle. The flank scales are from 0.5 to 1.0 mm wide, showing little differentiation from venter to midflank to dorsum. However, the larger scales occur towards the caudal region, and ventrally between the pre-pelvic and pectoral fins. The crown ornament is sometimes visible as an impression in the matrix, and one isolated broken fragment was cleaned in acid and cast in latex to show the crown surface of articulated scales (Fig. 4A). The crown ornament comprises 14–24 subparallel ridges or grooves extending to at least midcrown. The posterolateral crown edges may be denticulate (not well preserved). The scale neck is deep and strongly concave, and the base is shallow and convex. Some of the scales behind the head tesserae have been split through the middle showing a central red spot, interpreted as iron oxide infilling an enlarged primordial vascular space (sometimes preserved in acanthodian scales, e.g. Valiukevicius 1983, figs. 2.4, 3.4, 5). Scale patches on some of the loose flakes from ANU V773 were sectioned, but the histological structure is only poorly preserved; crown growth layers and "shadows" of the pore canals are sometimes discernible (Fig. 3F). Transverse natural sections through some scales cut through iron infilling of the pore canals low in the crown (Fig. 3E). Dark rounded spots in crown grooves on casts of the acid-cleaned impression of ANU V773 flank scales (Fig. 4A) possibly represent pore openings infilled with iron oxide. Scales identical to the normal body scales of ANU V773 are found in microvertebrate samples collected from Mount Crean (Fig. 6Q, R; see below). Natural sections of scales dorsal to the branchiostegal rays show vascular canals radiating through the posterior half of the scale (Fig. 3B); these vascular canals are not visible in scales further back on the body. The scale morphology of *Milesacanthus antarctica* gen. et sp. nov. thus corresponds closely with that based on a small number of scales from the original Mount Suess locality, referred to ?*Cheiracanthus* sp. and illustrated by Woodward (1921, pl. 1, figs. 12, 13).

Fin web scales are much smaller than flank scales, and decrease in size distally (Fig. 2D). The lateral line sensory canal is not discernible over most of the fish; apparently, it was not edged by specialised scales, at least as far as can be seen with the squamation preserved mainly as broken inner scale surfaces. Although scale patches of the caudal region are discernible on pieces C and D, overlying remnants of Fishes 2 and 3 obscure details of the squamation pattern on the tail.

Fins and spines

The preserved part of the anterior dorsal fin spine is 55 mm long, slightly curved, and inserted about 70 mm from the rostral point at an angle of 40° to the body. The inserted part is not visible, and possibly was prepared away. Five longitudinal ribs (probably excluding the leading edge rib) are exposed at its proximal end (Fig. 2D). The well-preserved fin web (adfw, Fig. 1A) extends to the tip of the spine. The posterior dorsal fin spine is relatively straight, deeply inserted in the body, and also makes an angle of 40° with the body surface. Its inserted length is 25 mm, and about 30 mm of the free end is preserved (partly as an impression) on piece C. The spine is preserved for its whole length of about 60 mm (mostly as

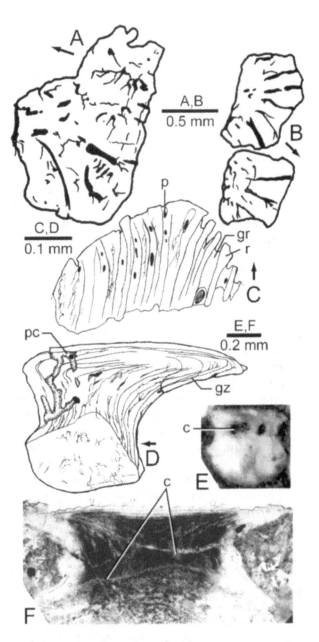

Fig. 3. Milesacanthus antarctica gen. et sp. nov. Scale histology of ANU V773 (A, B, E, F), and of isolated acid-etched Mount Crean scales (C, D). A: A natural horizontal section through head tessera dorsal to circumorbital bones (holotype; Fish 1). B: A natural horizontal section through the lower crown of two scales dorsal to the branchiostegal rays (holotype; Fish 1). C: A horizontal section through the anterior part of the crown of scale WAM 03.1.2 from Mount Crean. D: A vertical longitudinal section of scale WAM 03.1.1 from Mount Crean. E: A natural vertical transverse section through the front part of the crown, showing iron infilling of cross-cut canals low in the crown (scale from loose flake of ANU V773). F: A vertical transverse section, in front of the base apex, through a scale from a loose flake of ANU V773 (base partly missing), showing crown ridges, deep neck, many crown growth zones, and "shadows" of several canals low in the crown, some possibly showing canals rising up from them.

an impression) on the counterpart (pieces B and D). The remnants of the fin web preserved on piece C (pdfw, Fig. 1A) suggest that it also probably extended to the tip of the spine. Ventrally, the pectoral fin web base extends to 10 mm in front of the pelvic fin spine insertion. The two pectoral fin spines, presumably displaced on piece B (pectfs, Fig. 1B) are slightly curved, lack an insertion area, and have at least three smooth lateral ribs, decreasing in width to the posterior edge (not exposed), plus a wider leading edge rib. The more complete spine is 28 mm long with the tip missing. Their lack of insertion areas supports our interpreting them as pectoral fin spines, as this is a common condition when the spine is fixed to the scapulocoracoid rather than being only inserted in musculature (e.g. Denison 1979, fig. 20). The presumed admedian pectoral spine is ornamented with five ribs per side; the ribs converge at the apex of the spine. The total length of the pelvic spine is estimated at 30 mm. It has a relatively short insertion, is preserved at an angle of about 50° with the body, and again the fin web extends to the spine tip. This spine has three or four longitudinal ribs per side, plus a wider leading edge rib. The provenance of two other spine fragments in the pelvic region (sp, Fig. 1B) is undetermined. The anal fin spine (anfs, Fig. 1A–C) is deeply inserted at an angle of 60° to the body, with the insertion area 15 mm long and the total spine length estimated at about 50 mm. An ornament of at least five longitudinal ribs is preserved, possibly including a leading edge rib.

Fishes 2–4

Associated fish on ANU V773 seem to be the partial remains of three individuals of similar size to the holo-type, but considerably more disrupted and incomplete. Only the posterior dorsal fin spine (55–60 mm long) and part of the caudal region of Fish 2 are preserved. A natural section through the spine on the edge of piece C shows five ribs on each side plus a wider leading edge rib. Fish 2 overlies part of the caudal region of Fish 1 on the counter-part (pieces B and D). Fish 3 (Fig. 1B) is represented by a large patch of squamation and disarticulated fin spines surrounding a large element which is probably a sarcop-terygian scale. Fish 4 (Fig. 1B) comprises more disarticu-lated scale patches and fin spines, but it is also possible that these may be additional remains of Fish 3 which had "exploded" during decomposition. A dissociated ?pecto-ral fin spine 45 mm long is also preserved near Fishes 3 and 4, and various other scale patches are visible on bedding planes on the margins of the specimen. There is no evidence that these remains are not the same species as Fish 1 described above.

Isolated remains

Most of the isolated acanthodian remains from Portal Mountain and equivalent horizons at Mount Crean show similar morphology to the type material, and are there-fore provisionally included here. Some of these have been used to investigate the histology of spines and scales.

Of various isolated spines, ANU V775 (not figured) is an incomplete slightly curved example (about 35 mm long; maximum width about 3 mm) broken longitudi-nally to show the internal cavity. This is open posteriorly at the proximal preserved end, and has a cancellous tissue zone about 0.5 mm wide running along the length of its anterior border. Distally an apparent radiating structure within the spine is manifested in longitudinal section by fine lines of partly coalesced small pores, which seem to align with the midline of each rib, and with each groove. ANU V776 (Fig. 4B) is a straight spine (estimated original length about 60 mm) broken longitudinally through its right side. The insertion/exsertion boundary at the level of the broken section is 28 mm from the proximal end. The

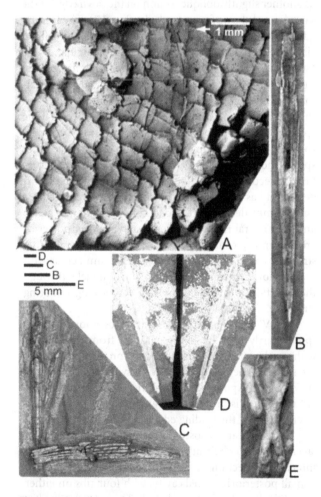

Fig. 4. A–D: *Milesacanthus antarctica* gen. et sp. nov.; specimens from Portal Mountain. A: A patch of squamation from a loose flake of ANU V773, prepared by etching in acetic acid to produce an impression (latex rubber cast whitened with ammonium chloride). B: An incomplete unpaired fin spine, trailing edge exposed (ANU V776). C: Two larger and one small very incomplete fin spines preserved in association (ANU V779). D: A fin spine and patches of squamation preserved in part and counterpart (AMF 55623). E: An isolated unpaired element, probably a ventral occipital ossification in dorsal view, with a small curved element (?branchial or circumorbital bone), possibly from *Milesacanthus antarctica* (Portal Mountain; ANU V774).

partly matrix-filled internal cavity is open at least as far back as 33 mm from the proximal end of the trailing edge, and surrounding very porous tissue extends as a core to the distal end of the preserved part. The outer zone shows thin discontinuous dark lines of fine connected pores, as in the previous specimen.

ANU V778 (spine length at least 60 mm; 4.5 mm thick at the proximal end of preserved tissue) shows a good broken section about 35 mm from the proximal end (Fig. 5A). The internal cavity is closed along the trailing edge, which is convex; the spine ornament comprises five longitudinal ribs on each side of a more pronounced leading rib. The tissue is quite vascular between the central cavity and the leading edge, with a radiating structure more evident anteriorly. The tip of the spine is preserved as another slightly oblique section on the sawn edge of the sample, and shows four lateral ribs (Fig. 5B). The leading edge rib and the core between the first pair of lateral ribs are composed of a lighter-coloured vascular tissue (stippled). A denser white tissue surrounds the central cavity, from which four or five canals radiate. The proximal end of this spine is preserved as an impression, and the first preserved tissue at the 18 mm level shows that the spine cavity is still open posteriorly, in section forming a rounded triangle 2 mm wide and 2 mm deep. Three other spines enclosed in the matrix in this sample are seen as broken sections. One about 12 mm from the spine just described has an almost circular section, 1.75 mm across and 1.5 mm deep with a broad convex leading edge rib and four lateral ribs on each side, and an open posterior groove and a centrally placed cavity with a radiating structure. A second large spine about 4 mm across with a posteriorly open cavity is broken obliquely through the insertion area, which shows a very strong radiating structure.

ANU V779 (Fig. 4C) shows an association of three spines partly exposed by preparation from the matrix, with another spine revealed in section on the underside of the sample. The best preserved is a symmetrical, slightly curved spine exposed in anterior view. The insertion area (14 mm long) is broken through anteriorly to show the internal cavity, the leading edge rib is 1.5 mm wide, and two lateral ribs are exposed on each side. The broken distal end protrudes beneath the second, less well-preserved, spine, and in section shows a broad posterior groove and a small posteriorly placed cavity, with four ribs on either side of the broader leading edge rib. The second spine has 39 mm of the proximal end preserved, an insertion area 14 mm long, and five ribs exposed on either side of the leading edge rib (poorly preserved). The central cavity is open along the trailing edge to at least 25 mm from the proximal end. These two similar spines could be anterior and posterior dorsal spines from one fish. AMF 55623 (Fig. 4D) is an almost complete unpaired spine about 60 mm long surrounded by a patch of indeterminate

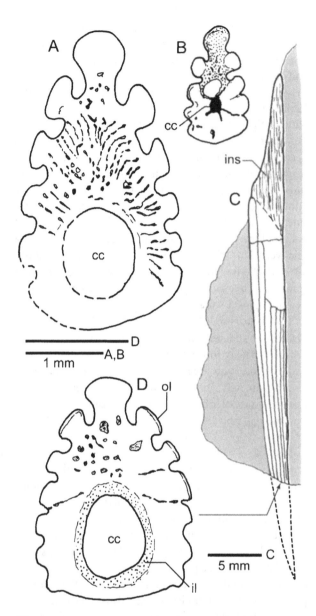

Fig. 5. *Milesacanthus antarctica* gen. et sp. nov.; fin spines and transverse sections. A, B: Natural transverse sections on an isolated spine (Portal Mountain; ANU V778): A: A broken transverse section at 35 mm from the proximal end. B: A sawn slightly oblique section from near the tip of the spine. C, D: An isolated spine (Mount Crean; ANU V880). C: As preserved. D: A broken section at the distal preserved end.

scales, which is preserved in part and counterpart, split through the middle to show the central cavity.

ANU V860 (higher horizon at Portal; not figured) displays a small broken spine fragment 1 mm across showing three longitudinal ribs. ANU V1178 (not figured; length 30 mm) is a part impression of an incomplete spine showing four longitudinal ribs. ANU V2165 is the subcircular end of a similar ribbed spine, weathered out around the matrix-filled central cavity. ANU V2174 is a straight ribbed spine at least 85 mm long (maximum width about 4.5 mm), broken longitudinally to show that the central cavity is open posteriorly about 45 mm from the proximal end.

ANU V2163 (Fig. 6H, O, P) has been sectioned at two levels near the proximal end of a spine with a rounded cross-section. Vascular/dentine canals radiate from a circular central cavity, and also cut through some longitudinal canals above the central cavity. The spine has a broader leading edge rib, and four lateral ribs of which those closer to the leading edge are widest (Fig. 6H). The sections show a superficial layer on the "crest" of the ribs, which thins and disappears in the grooves. This outer layer is penetrated by very fine, branching dentine tubules rising from the underlying tissue which forms the base of the ribs (Fig. 6P). The latter tissue has branching dentine tubules radiating from a network of vascular/dentine canals, with only rare lacunal widenings and without any visible bone cell lacunae, comparable with the structure as described by Woodward (1921, p. 55) in *Byssacanthoides*. The radiating canals are surrounded by a thin layer which is whiter than the interdenteonal matrix, with only short tubules extending from them. Again, no bone cell lacunae are detected. Close to the central cavity, the canals are aligned either more or less parallel transversely, or longitudinal, to the inner surface. The sections show no evidence of an inner bone layer lining the central cavity. These spine sections are more rounded than other examples (Fig. 5A, B, D), and this presumably reflects the shape variation in different spines of one fish. However, the relatively deep and sometimes undercut grooves, separating the clearly defined slightly convex flat-topped ribs, are consistent differences between these Antarctic spines and spines with similar histology from Europe (e.g. *Onchus overathensis*, *Devononchus concinnus*; Gross 1933a, b).

Similar features are observed in the Mount Crean material. Unfigured material includes ANU V890, a straight ribbed spine (1.5 mm wide; 28 mm of proximal end exposed) showing the finely striated inserted portion, and ANU V968, two small straight spines (preserved length 14 mm, 1.5 mm wide), one broken to show the internal cavity extending right to the tip. ANU V969 is a small spine fragment, and AMF 55859 was a large straight (probably unpaired) spine, of which an incomplete impression 27 mm long and 5 mm wide is preserved. AMF 55886 shows several associated incomplete spines up to 40 mm long, and AMF 55873 shows a scattering of scales and spines, the most complete with an exserted portion 51 mm long (missing the tip) and an inserted portion 16 mm long, and preserved in outline to show the tapered insertion.

More complete is ANU V880, a relatively straight, apparently unpaired spine missing the distal end (Fig. 5C; estimated length 45 mm). The inserted portion shows what we term the "tapered" condition, rapidly decreasing in width just past the insertion/exsertion boundary

(visible on the opposite side of this specimen). This is probably a dorsal fin spine, by comparison with the holotype. The broken tissue at the proximal end exposes the impression of the internal cavity, open along the trailing edge, but how far back it remained open is obscured by matrix. One leading edge and six lateral ribs are seen on the exposed right side. A natural cross-section at the preserved extremity (Fig. 5D) shows that the lateral ribs decrease in width posteriorly, a distinctly denser basal layer lines the spine central cavity, and some of the anterior ribs show a dense superficial layer under oil of anise, as in the sections of Portal spines (Fig. 6H). The trailing edge is gently convex from side to side, and the central cavity is subcircular. The tissue in front of the cavity is vascular, with a few larger vacuities, but no subcostal canal. Some radiating internal structure is suggested by fractures in ribs 3–4. Overall this spine is similar to the reconstruction of the dorsal spine of *Onchus overathensis* by Gross (1937, fig. 29), and that form also has a radiating structure in cross-section (Gross 1933a, fig. 11), but with a much more triangular section. The cross-section shape of other spines (Fig. 5A, B, D) is closer to *Devononchus concinnus* (Gross 1933b, fig. 6). The main differences shown by the latter species are the shallower grooves between ribs and the conspicuous subcostal canal.

Scales identical to the normal body scales on the holotype of *Milesacanthus antarctica* are also found at Mount Crean. ANU V893 (not figured) shows the external surface of a typical scale, almost 1 mm across, with 14 subparallel ridges on the anterior half of the crown. In all preserved respects it is almost identical to that figured by Woodward (1921, pl. 1, fig. 12). The same scale type, showing pore canal openings between the anterior ends of crown ridges, has been illustrated from Mount Fleming (Young 1989a, fig. 4C). Other very similar scales from acid-prepared microvertebrate samples show the deep concave neck and convex base (Fig. 6Q, R), and again are very similar to the scale cross-section originally figured by Woodward (1921, pl. 1, fig. 13). In these acid-prepared scales, the histology is much better preserved than in the holotype. They show thin, distally branching dentine tubules and canals rising through the neck and curving over into the upper parts of the growth zones of the crown. They also have a pore canal system penetrating the anterior part of the crown, with canals opening out through pores in the grooves between the ribs on the anterior half of the crown and in the lower neck (Fig. 3C, D). No bone cell lacunae are present in the crown or base; the latter is penetrated by Sharpey's fibres and thin, irregularly branching canals.

Finally, ANU V774 (Fig. 4E) is an isolated unornamented element almost 12 mm long, and probably not a dermal bone. The "mandibular ossifications" (possible ceratohyals) in *Diplacanthus horridus* (Gagnier 1996, fig. 7) have a similar expanded end, but ANU V774 is clearly a symmetrical median element, which in shape compares with the ventral occipital ossification of *Acanthodes bronni* (Miles 1973b, fig. 5). If it is this element, it is more elongate than in that form, with a narrower central region. The posterior expanded part resembles *Acanthodes bronni* with its rounded lateral ridges; the anterior expansion is more elliptical than in *Acanthodes bronni*, which has more prominent anterolateral corners (Miles 1973b, fig. 11A, pl. 1A). The anterior and posterior borders of ANU V774 are broken, so it is not clear how much is missing. A contiguous small bone could be a branchial or circumorbital bone, and scattered indeterminate scales are preserved in the matrix.

Fin spine structure and histology

From the evidence of both isolated material and articulated specimens, we can characterise the fin spines of *Milesacanthus antarctica* gen. et sp. nov. as follows: the pulp canal was open for most of the spine length, only being enclosed along the distal quarter or third; closely spaced, narrow vascular/dentine canals radiate out from the central cavity towards the external surface (approximately 8 canals/mm); the spines lack an enlarged subcostal canal, with the longitudinal vascular/dentine canals less than 0.1 mm wide, with narrow denteons formed round them; finally, the inner, presumed osseous, layer is thin and relatively dense.

The similar microstructure between *Byssacanthoides* fin spines and *Devononchus concinnus* (Gross, 1930) was noted by White (1968, p. 12), and this is also evident in the new material of *Milesacanthus antarctica* described above. *Devononchus* spines have three to five smooth longitudinal ribs per side, the insertion area of unpaired fin spines tapers rapidly just proximal to the insertion /exsertion boundary, and the histological structure is also similar (Gross 1933b, fig. 6). *Milesacanthus* fin spines differ in having a more rounded than oval cross-section, and lacking a subcostal longitudinal canal (also seen in *Diplacanthus* fin spines; Gagnier 1996, pp. 153, 154). Also, the pulp cavity is only open for a third of the length along the trailing edge in *Devononchus concinnus* spines (cf. Lyarskaya 1975, who followed other Russian workers in reassigning the species to *Archaeacanthus concinnus*). By comparison with other acanthodian taxa based on whole fish, the "tapered" insertion area might only apply to the anterior dorsal fin spine (cf. *Diplacanthus striatus* in Watson 1937, figs. 14A, 15).

Scale structure and histology

To summarise scale structure, in *Milesacanthus antarctica* scale crowns are ornamented with 14–24 subparallel ribs or grooves extending to the midcrown or further posterior; posterolateral crown edges are sometimes denticulate. The scale neck is deep and strongly concave, and the base is shallow and convex. Sections of Mount Crean

scales show thin dentine tubules and canals in the crown, but also have a pore canal system through the anterior crown, with canals opening out through pores in the grooves between the ribs on the anterior half of the crown and in the neck (Fig. 3C, D).

In histological structure, *Milesacanthus* scales resemble those of *Haplacanthus perseensis* (Frasnian, Latvia) in having long dentine tubules rising up and curving over into the upper growth zones of the crown (Denison 1979, figs. 9I, 10L), although the latter lack a crown pore canal system. Other non-poracanthodid acanthodians which have a pore canal system in the scale crown include *Ptychodictyon*, *Ectopacanthus? pusillus*, *Lietuvacanthus*, and sometimes *Nostolepis gracilis*. Of these, *Nostolepis gracilis* scales are most similar to the Antarctic scales, in having pores opening out between ribs on the anterior part of most scale crowns (Gross 1971; Valiukevicius 1998; Vergoossen 1999, pl. 6.7). However, this taxon is much older (Late Silurian–earliest Devonian), with only a few widely spaced ribs extending the whole length of the crown, and nostolepid-type histology, with bone cell lacunae throughout the base and crown, and Stranggewebe (a specialised mesodentine formed from parallel elongated lacunae/tubules) in the crown. Scales of *Ectopacanthus? pusillus* Valiukevicius, 1998 also resemble those of *Milesacanthus* in having a pore canal system in the anterior crown, but all figured scales of *Ectopacanthus? pusillus* are very worn, and the canals only open through pores onto the anterior neck, and not on the crown; this taxon is also considerably older (Lochkovian Stoniskiai Formation, Lithuania). *Ptychodictyon* is the only one of these taxa assigned to the Diplacanthidae, but its scales have a distinctive "micro"pore canal system confined to each growth zone (Gross 1973, figs. 10, 11). Scales very similar externally to those of *Milesacanthus* are also found in the Emsian Jauf Formation of Saudi Arabia (Lelièvre *et al.* 1994, fig. 2.15).

Whole fish reconstruction

A trace of the outline of the holotype was used for the reconstruction in Fig. 1C, although the shape of the caudal region is uncertain. This reconstruction summarises the general features of *Milesacanthus antarctica* gen. et sp. nov. as described above. It was less deep-bodied, with more triangular fin webs, and less robust fin spines when compared with *Culmacanthus stewarti* (Long 1983, fig. 9), and lacked the enlarged cheek plate and circumorbital bone characteristic of *Culmacanthus* and *Diplacanthus striatus*. The sensory line canals, as far as preserved, show a standard pattern. Only post-orbital elongated circumorbital plates were present, with the largest carrying a sensory line. Anterior plates were presumably absent. By comparison, in *Diplacanthus* the

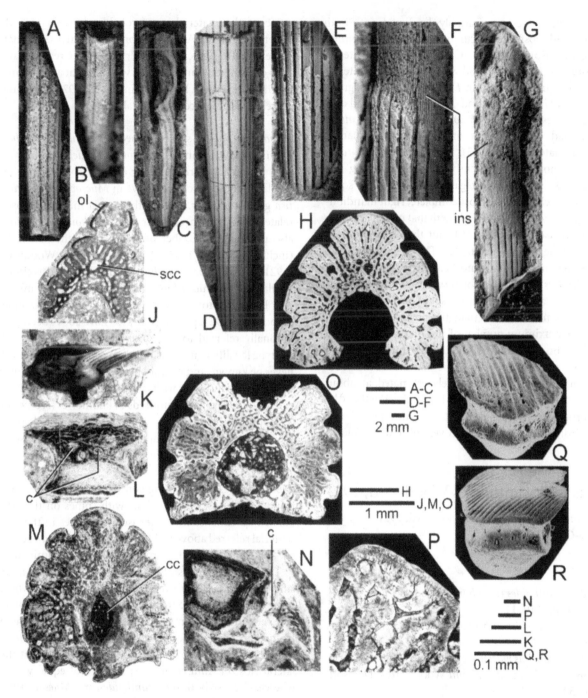

Fig. 6. A–C: *Byssacanthoides debenhami* Woodward, 1921. A latex cast of lectotype NHM P12553 (Granite Harbour), provided by the Natural History Museum, London, in anterolateral, anterior, and trailing edge views, respectively (irregular margin in C is an artefact). D–F: *Antarctonchus glacialis* White, 1968. D: A latex cast of holotype NHM P49164 (Boomerang Range) in anterior view (middle part of spine), showing the broader leading edge rib; E, F: A cast of spine GS 7395/15 (Boomerang Range). E: Lateral view of exserted portion. F: Insertion/exsertion boundary. G: ?*Antarctonchus* spine AMF 55549 (Mount Fleming), showing insertion/exsertion boundary. H: *Milesacanthus antarctica* gen. et sp. nov., fin spine ANU V2163, transverse section (Portal Mountain). J, K: ?*Culmacanthus antarctica* Young, 1989b (ANU V882, Mount Crean). J: A fin spine transverse section. K: A flank scale vertical longitudinal section. L–N: ?*Culmacanthus antarctica* Young, 1989b (ANU V970, Mount Crean). L: A vertical transverse section of scale showing wide canals low in the crown. M: A fin spine transverse section. N: Two scales near the spine section, the vertical transverse section on the left, and off-centre a longitudinal section showing wide canals in the anterior crown on the right. O, P: *Milesacanthus antarctica* gen. et sp. nov., fin spine ANU V2163 (Portal Mountain). O: A transverse section (more distal than in H). P: Detail of one rib from the upper right side. Q, R: *Milesacanthus antarctica* gen. et sp. nov.; two isolated acid-etched scales (Mount Crean locality MC3, GCY SEM stub 94/27/3,4). (A–G are latex rubber casts whitened with ammonium chloride; the proximal end of spine fragments is uppermost).

circumorbital plates were variably developed, with both anterior and posterior circumorbitals in *Diplacanthus striatus* (Denison 1979, fig. 4C), only anterior plates in *Diplacanthus horridus* Woodward, 1892, and anterior and ventral circumorbitals in *Diplacanthus ellsi* Gagnier, 1996. *Tetanopsyrus lindoei* and *T. breviacanthias* had only anterior plates (Hanke *et al.* 2001), while *Culmacanthus stewarti* had both anterior and the greatly enlarged posterior plates (Long 1983). *Milesacanthus antarctica* gen. et sp. nov. was apparently unusual in having two pairs of close-set branchiostegal rays separated by squamation on each side. The most complete, right mandibular splint is comparable in shape with that of *Diplacanthus striatus* (Watson 1937, fig. 15), but the elements seem relatively long in the new taxon. The scapulocoracoid, with its high, narrow shaft, most closely resembles that of *Culmacanthus stewarti*. As for most non-climatiid acanthodians, the anterior dorsal fin spine on *Milesacanthus antarctica* is more curved than the posterior dorsal fin spine. The maximum number of ribs on fin spines of the holotype is no more than 11, and the maximum estimated length is about 75 mm for the anterior dorsal fin spine (inserted part conservatively estimated at about one-quarter the total length in Fig. 1C). Whether the reconstructed procoracoid might be re-interpreted as a pelvic girdle must await the discovery of new articulated material.

Genus Byssacanthoides Woodward, 1921

Byssacanthoides debenhami Woodward, 1921
Fig. 6A–C

Synonymy. –

1921 *Byssacanthoides debenhami* Woodward, pp. 54, 55, pl. 1, figs. 10, 11
1968 *Byssacanthoides debenhami* Woodward – White, p. 12, fig. 2
1968 *Antarctonchus glacialis* – White, pp. 11, 12 *in pars*
1979 *Byssacanthoides debenhami* Woodward – Denison, p. 50, fig. 33J

Type specimens. – Three spine fragments (NHM P12553–555), of which the largest (P12553) was selected as the lectotype by White (1968, p. 12).

Other material. – Possibly GS7398/12, tentatively assigned by White (1968) to *Antarctonchus*.

Locality and horizon. – The type material came from glacial moraine at Gondola Ridge, Mount Suess (locality 2 of Young 1988, fig. 3). It is of unknown provenance, but associated turiniid thelodont scales suggest that at least some of the material was probably derived from the lower beds of the Aztec Siltstone. Specimen GS7398/12 came from the higher fossil layer at Mount Crean (locality MS5 of Gunn & Warren 1962), which might approximate to Unit 8, Section L2 of Askin *et al.* (1971), although correlations are uncertain (see above).

Remarks. – The fin spines named *Byssacanthoides debenhami* Woodward and *Antarctonchus glacialis* White are here retained as form taxa for isolated spines. They are similar to those described above in *Milesacanthus antarctica* gen. et sp. nov., so the three taxa might be closely related. The type locality for *Byssacanthoides debenhami* also produced a number of isolated acanthodian scales (including NHM P12559, 12576), named by Woodward (1921, p. 56) as *Cheiracanthus* sp., and described as closely similar to *Cheiracanthus murchisoni* from the middle Old Red Sandstone of Scotland. The figured examples (Woodward 1921, pl. 1, figs. 12, 13) have above been provisionally referred to *Milesacanthus*. The problem with the type locality material is that its original derivation will never be known, and it could have come from a variety of horizons and localities within the Aztec Siltstone, perhaps extending over tens or hundreds of kilometres (see Young 1988, pp. 10, 11, 22). Thus, remains within a single piece are the only ones that can be attributed to the same (unknown) original locality.

Description. – The following comments are based mainly on a comparison of spine casts with spines on the articulated specimens, and other isolated spines in the new material referred above to *Milesacanthus antarctica* gen. et sp. nov. Woodward's type material was also examined.

The type specimens of *Byssacanthoides debenhami* are small spine fragments which, by comparison with whole spines, probably came from the distal third of the spine. The very limited type material presents few distinguishing features from the much larger fin spines named *Antarctonchus glacialis* White (see below), but it seems that the lateral ribs on *Antarctonchus* spines are of equal width, whereas those of both *Byssacanthoides*, and *Milesacanthus* described above, are graduated or variable. In the lectotype of *Byssacanthoides debenhami* (Fig. 6A–C), the leading edge rib makes up more than half the width of the spine in the anterior view, but a rib of this width has not been encountered in any of the material referred above to *Milesacanthus*. The spine sections (Woodward 1921, pl. 1, fig. 11a; White 1968, fig. 2c), presumed to come from the distal third, are also much more rounded than the equivalent distal parts of *Milesacanthus* spines described above (e.g. Fig. 5D). For these reasons *Byssacanthoides debenhami* has been retained as a separate form taxon for isolated spines.

Genus Antarctonchus White, 1968

Antarctonchus glacialis White, 1968
Fig. 6D–G

Synonymy. –

1968   *Antarctonchus glacialis* White, pp. 11, 12, pl. 2, fig. 3 *in pars*

?1972   *Antarctonchus* sp. – Ritchie, p. 351 *in pars*

1979   *Antarctonchus glacialis* White – Denison, p. 50, fig. 32G

?1991   "incomplete fin spine" – Young, fig. 15.7(a)

Holotype. – NHM P49164, possibly including GS 7395/8, 13, 18, which might be part of the same spine.

Other material. – NHM P49166, GS7395/7, 14, 15, 16 from the type locality; ?AMF 55549 from Mount Fleming. *Localities and horizon.* – The type locality is MS2 of Gunn & Warren (1962), in the Boomerang Range (Young 1988, fig. 3, locality 20), from a horizon probably some 70 m above the base of the Aztec Siltstone (Young 1988, p. 9). The Mount Fleming specimen is only provisionally included. It comes from locality MS228, the lowest fossil-iferous horizon in Section 26 of Barrett & Webb (1973), in the upper beds of the Beacon Heights Orthoquartzite which conformably underlies the Aztec Siltstone (Young 1988, fig. 3, locality 4).

Remarks. – White (1968) relied on four characters to differentiate his new taxon from *Byssacanthoides debenhami* Woodward. In the latter the spines are smaller, with ribs graduated, ribs few in number, and with a "wide" rather than "narrow" posterior area. As the original *Byssacanthoides debenhami* specimens are probably distal fragments, White's "size" criterion can be discounted. The form of the posterior face, or trailing edge, also varies with distance from the tip, as seen in various fin spines described above. The spines in the *Antarctonchus* type material are considerably larger than any other specimens described above, so the maximum attained size might be a valid criterion, even though the greater number of ribs for *Antarctonchus* is probably also size related. In acanthodians in general, the number of spine ribs appears to increase with the age of the fish, and on individual spines, the number increases proximally. As none of the type *Antarctonchus* spine fragments shows the structure of the trailing edge distal to the inserted area, the descriptions by White (1968, p. 11) of a "smooth flattened area" and Denison (1979, p. 50) of "a narrow posterior channel" can be discounted.

Comparing casts of the spines of *Antarctonchus glacialis* White with those of *Byssacanthoides debenhami* Woodward, and the extensive material including fin spines described above as *Milesacanthus antarctica* gen. et sp. nov., provides some support for White's (1968) distinguishing criteria, and *Antarctonchus glacialis* is retained here as a form taxon for isolated spines, normally of large size, with smooth longitudinal ribs, a wider leading edge rib, and up to eight lateral ribs which are of equal width.

Description. – The original *Antarctonchus* fin spines were preserved only as impressions, so their histology is unknown, although White (1968) noted that the spines had a large internal cavity. Latex casts of the fin spine impressions of both *Antarctonchus* and *Byssacanthoides* in the London and Wellington collections have been compared to clarify their structure (Fig. 6A–F). The *Antarctonchus glacialis* holotype (NHM P49164) is 87 mm long as preserved (White 1968, pl. II, fig. 3) but lacks the proximal part and the trailing edge. White (1968, p. 11) described it as "rounded in cross-section with very little lateral compression", but latex casts (Fig. 6D) indicate that "subtriangular in section" is a more accurate description. *Antarctonchus glacialis* specimen GS 7395/15 (Fig. 6E, F) was described by White (1968, p. 11) as having an insertion "2.7 cm long in front and runs backwards and upwards at an angle of 45°, which gives the backward slope of the spine" (i.e. the spine was inserted in the body at an angle of 45°). White's (1968, p. 11) observation that the "anterior median" (leading edge) rib was not preserved is incorrect, because the left preserved edge in Fig. 6E includes part of the leading edge rib. Thus, this spine had 17 ribs (eight on each side), not the 19 ascribed by White. White's (1968, p. 11) statement that the "roughened inserted area is continued as a narrow strip in the middle of the posterior face of the spine" is misleading, as the "posterior face" (trailing edge) is wide open, with the open central cavity edged by the side walls of the spine.

As noted by White (1968), *Antarctonchus glacialis* specimen GS 7395/8+/13+/18 (erroneously listed as /19 by White) also shows that the large central cavity is wide open for the length of the insertion area, and for some distance along the exserted part. White (1968) suggested that the NHM holotype and these GS specimens could be part of the same spine, and this is supported by restudying the specimens (showing that trimmed latex rubber casts fit reasonably well together). This evidence suggests a total spine length of some 14 cm, with the insertion area 3 cm long (along the leading edge), and ornamented with a maximum of 17 longitudinal ribs. White (1968, p. 11) also noted that the remaining spine fragments might have been originally associated, so the collected type material could represent incomplete remains of spines belonging to one fish.

AMF 55549 (Fig. 6G) from the Beacon Heights Orthoquartzite at Mount Fleming is the oldest acanthodian spine known from the Beacon sequence. It is an impression of a large spine fragment with at least seven equal-width ribs per side. The incompletely preserved insertion area is approximately 30 mm long, suggesting a total spine length of over 12 cm. The insertion/exsertion boundary is almost perpendicular to the longitudinal axis, so the spine would probably have protruded from the body at a much higher angle than any of the other spines dealt with above. Ritchie (1972) referred AMF 55549 to *Antarctonchus* sp., which it resembles more than *Milesacanthus* in its large size, equal-width ribs, and similarly developed insertion (ins, Fig. 6F, G). However, this spine comes from much lower in the sequence, so Ritchie's provisional assignment is followed here with great reservation.

Family Culmacanthidae Long, 1983

Genus Culmacanthus Long, 1983

Culmacanthus antarctica Young, 1989b
Fig. 6J–N

Synonymy. –

1988   "acanthodian spines and scales"; "acanthodians" – Young, p. 12 *in pars*
1989a *Culmacanthus antarctica* – Young, p. 51
1989b *Culmacanthus antarctica* Young, pp. 14–17, figs. 2A, 3A
1992   "acanthodians" – Turner & Young, p. 90
1993   *Culmacanthus antarctica* Young – Young *et al.*, p. 248

Holotype. – ANU V967 (CPC 26579), a right cheek plate.

Other material. – ANU V970, an isolated fin spine with associated scales, is referred to this species. ANU V882 (spine and scale thin section) is very tentatively included.

Locality and horizon. – Mount Crean, Lashly Range, locality 8 (Young 1988, figs. 3, 4); collecting sites MC1 (V882) and MC3 (V970), from lower units in the Aztec Siltstone (Section L2 of Askin *et al.* 1971).

Remarks. – *Culmacanthus antarctica* Young, 1989b was erected for a single distinctive cheek plate differentiated from the type species *Culmacanthus stewarti* Long, 1983 by the position of the sensory groove passing off the ventral margin, and its shape and proportions. It was recognised that some of the isolated "*Antarctonchus*" spines in the Aztec fauna could belong to *Culmacanthus antarctica* (Young 1989b, p. 14), but without the comprehensive study of acanthodian material presented here, no criteria for distinguishing such spines were established. The holotype of *Culmacanthus antarctica* came from the

same locality and horizon as the spine described below, which is now recognised to show several distinctive features of spines on articulated examples of the type species *Culmacanthus stewarti*. Given that some semi-articulated fishes (palaeoniscoids, sarcopterygians) are known from Mount Crean, it is possible that new specimens might be found to corroborate that this spine belongs to *Culmacanthus antarctica*.

Description. – *Culmacanthus stewarti* Long, 1983 from the Givetian–Frasnian of Victoria is represented by various more or less complete articulated specimens displaying a range of fin spines, and we first summarise the known fin spine morphology. Long (1983, p. 59) described the fin spines as "ornamented with approximately nine coarse ribs, four being visible on each lateral face of the spines and two narrower ribs present on the posterior face between the larger ribs which form the posterolateral margin". The spine cross-section was described as relatively narrow with a flat posterior face. The largest fin spines of *Culmacanthus stewarti* are about 6 cm long, with a maximum of about nine longitudinal ribs. The type material is preserved only as impressions, so spine histology is unknown. Long (1983) regarded the fin spines as possibly synonymous with very similar isolated spines named *Striacanthus sicaeformis* Hills, 1931 from the Upper Devonian of Victoria. These are up to 35 mm long, with four or five longitudinal ribs per side. Their histology is known only to the extent that Hills (1931, p. 214) described "a central longitudinal pulp cavity with smaller tubes arranged concentrically around it parallel to its length". A longitudinal canal structure is also demonstrated in older acanthodian spines from Australia referred to "*Striacanthus*" (Burrow 2002, fig. 14L). This seems rather dissimilar to the distinctive radiating structure of canals in all the Antarctic spines so far investigated, so on this evidence *Striacanthus* and *Culmacanthus* might not be synonymous, and Long's (1983) suggestion of retaining *Striacanthus* as a form taxon for isolated spines of *Culmacanthus* would not be valid.

The articulated type specimens of *Culmacanthus stewarti* show the fin spines mainly in lateral view. It is unclear whether the leading edge rib was wider than the lateral ribs, but a cast of NMV P159838 in the ANU collection shows that one of the four figured ribs on the "pectoral spine" of Long (1983, fig. 3) is a slightly wider and more prominent leading edge rib; on other pectoral fin spines, the ribs that can be seen appear to be of equal width (Long 1983, fig. 4D). We note that the long striated base of insertion shown by Long (1983, fig. 3) is anomalous for a diplacanthid pectoral spine, and if correctly identified would mean a deep articulation with the scapulocoracoid, and thus significantly different to the superficial connection of the pectoral spine in *Diplacanthus* (Hanke *et al.* 2001, p. 752). This aspect of *Culmacanthus* needs to be reinvestigated.

The dorsal fin spines of *Culmacanthus stewarti* clearly differ from those of *Byssacanthoides*, *Antarctonchus* and *Milesacanthus* in having a very wide lateral rib towards the trailing edge (Long 1983, fig. 1A). This feature would be difficult to establish in some isolated Antarctic spines that are split longitudinally, but is demonstrated in the one specimen (ANU V970) referred here to *Culmacanthus antarctica*. This assumes that this aspect of fin spine morphology is a generic character for *Culmacanthus*. ANU V970 comes from the same locality as the holotype of *Culmacanthus antarctica* (ANU V967), and the matrix has identical lithology.

ANU V970 is a straight spine about 70 mm long with the proximal 42 mm exposed, and a row of scales along its length. The spine cavity is open posteriorly to at least 36 mm from the proximal end. The cross-section shape is revealed by a section cut about 20 mm from the tip (Fig. 6M). This shows narrow canals radiating out from the central cavity, as in the other spine sections described above, but with some significant differences. The central cavity is laterally compressed, rather than round as in other spines (e.g. Figs. 5A, D, 6O). The more complete left side of the section shows a broader leading edge rib, two small ribs forming the posterolateral corners, and three main lateral ribs increasing in width towards the posterior. This is the reverse of the situation in *Byssacanthoides* or *Milesacanthus*, where the lateral ribs decrease in width posteriorly (Figs. 5A, B, D, 6H). The correspondence with the "two narrower ribs present on the posterior face between the larger ribs which form the posterolateral margin" in Long's (1983, p. 59) fin spine description is unlikely to be coincidental. The central cavity in ANU V970 (cc, Fig. 6M) is lined by dense, presumably osseous, lamellae; there is no subcostal canal. A thin superficial layer covering the "crest" of the ribs is clearly differentiated from the underlying tissue.

The histology of some scales associated with this spine is also shown in the thin section of ANU V970, although there is no good evidence on whether they belong with the spine or not. One vertical transverse section cuts through the ridges ornamenting the anterior part of the crown, and through several canals low in the crown (c, Fig. 6L). Of two scales adjacent to the fin spine, one is another vertical transverse section, probably through the posterior half (as no canals are intersected), while the other is an oblique longitudinal section which cuts through one of the pore canal openings between the crown ridges (Fig. 6N). Scale histology is unknown in the type species, but on the basis of a good cast of scale impressions, Long (1983, fig. 8A) suggested that *Culmacanthus* scales had a pore canal system. However, there is no clear evidence of pore openings on the scale crowns of *Culmacanthus stewarti* (but they are poorly preserved), and the wide canals exposed are possibly vascular canals leading back under the grooves or ridges.

A second spine section (Fig. 6J) is very provisionally included here, mainly because it differs from the spines ascribed above to *Byssacanthoides*, *Antarctonchus* or *Milesacanthus*. It is also different to ANU V970 just described, but the holotype of *Culmacanthus stewarti* demonstrates that the dorsal fin spines differ in rib morphology from the other spines (Long 1983, fig. 1), so the same situation might apply for *Culmacanthus antarctica*.

ANU V882 from Mount Crean shows a thin section evidently from the proximal part of a spine, with the trailing edge wide open. This spine was more laterally compressed and triangular in cross-section compared with the Portal Mountain spines, and clearly resembles sections of *Onchus overathensis* (e.g. Denison 1979, fig. 33B), in which similar radiating striations are developed (Gross 1933a, fig. 11). The section cuts through several enlarged longitudinal canals, of which the largest (scc, Fig. 6L) is comparable in position and size to the subcostal canal that typifies *Devononchus concinnus*, which also has strong radiating striations (e.g. Gross 1933b, fig. 6). This is the only Antarctic spine investigated that shows a subcostal canal, but it differs from *Devononchus concinnus* spine sections in other respects. ANU V882 is more triangular, has more, and better defined, ribs, and deeper intervening grooves, and also shows a clearly differentiated superficial layer (ol, Fig. 6J). In *Diplacanthus* the fin spines are also described as having "two pulp cavities of which the posterior is the larger" (Gagnier 1996, pp. 153, 154), but whether this is variable within a taxon, as implied by the two different spine sections referred here to *Culmacanthus antarctica*, must await a histological study of all spines in an articulated specimen.

An associated scale in ANU V882 (Fig. 6K), preserved as a vertical longitudinal thin section next to the spine, differs from those of *Milesacanthus* described above in having a lower neck and a more convex base, although some scale impressions illustrated for *Culmacanthus stewarti* are a little different (Long 1983, fig. 8B). *Culmacanthus* scales were described as deep-based with a constricted neck and a relatively flat crown, and the main difference from the scales of *Milesacanthus antarctica* described above is that they have only six or seven crown ridges (Long 1983). Again, the extent to which scale morphology is variable within a taxon can only be determined for taxa based on articulated fish, and *Gladiobranchus probaton* (Lower Devonian, Canada), for example, has at least two scale types, with smooth-crowned scales on the posterior flank of the body, and *Nostolepis gracilis*-type scales on the fin webs, mid-dorsal line, and ventrally (C.J. Burrow, pers. obs.). Whether the latter type of scales shows pore openings in the grooves, as described above for *Milesacanthus*, is not known. As noted above, the potential exists for finding more semi-articulated fishes from the *Culmacanthus antarctica* type locality at Mount Crean, so a new specimen might resolve some of these questions for this taxon.

Acknowledgements

Drs P. J. Barrett and A. Ritchie are thanked for arranging participation in the Victoria University of Wellington Antarctic Expedition under the New Zealand Antarctic Research Program (GCY). Field and logistic support was provided by the 1970–1971 staff at Scott Base, US Navy Squadron VXE-6, and other members of the VUWAE 15 team (R. Askin, P. Barrett, R. Grapes, B. Kohn, J. McPherson, D. Reid, A. Ritchie). GCY also acknowledges travel grants to London in May 1988 by the Trans-Antarctic Association, and to New Zealand in 1985, for study of comparative material. Dr A. Ritchie (Australian Museum), Dr J. A. Long (Western Australian Museum), and Dr Craig Jones (Institute of Geological and Nuclear Sciences, Wellington) provided information and/or loan of Antarctic material in their respective institutions. Sally Young provided latex casts of type material held in the Natural History Museum, London, and Alex Ritchie prepared some material, and provided digital images of some AM specimens. Drs R. E. Barwick and J. Caton assisted with scanning and computer preparation of illustrations, and B. Harrold provided essential computer support at ANU. GCY acknowledges financial support in Canberra by ANU Faculties Research Fund Grants F01083 and F02059, and overseas by the Alexander von Humboldt Foundation, for a Humboldt Award in Berlin (2000–2001). Dr P. De Deckker (ANU) and Professor H.-P. Schultze (Museum für Naturkunde, Berlin) are thanked for the provision of facilities in their respective departments. CJB acknowledges support of an ARC Discovery Project and Postdoctoral Research Fellowship (2002–2004) and the facilities provided by the Department of Zoology and Entomology, University of Queensland. We wish to thank reviewers Drs Juozas Valiukevicius and Pierre-Yves Gagnier for their helpful comments. This research was a contribution to IGCP Projects 328, 406, 410, and 491.

References

Agassiz, L. 1844–1845: *Monographie de poissons fossiles des Vieux Grès Rouges ou Système Dévonien (Old Red Sandstone) des Îles Britanniques et de Russie*. Neuchâtel.

Askin, R.A., Barrett, P.J., Kohn, B.P. & McPherson, J.G. 1971: Stratigraphic sections of the Beacon Supergroup (Devonian and older (?) to Jurassic) in South Victoria Land. *Publication of the Geology Department, Victoria University of Wellington, Antarctic Data Series 2*, 1–88.

Barrett, P.J. & Webb, P.N. (eds) 1973: Stratigraphic sections of the Beacon Supergroup (Devonian and older (?) to Jurassic) in South Victoria Land. *Publication of the Geology Department, Victoria University of Wellington, Antarctic Data Series 3*, 1–165.

Berg, L.S. 1940: Classification of fishes, both Recent and fossil. *Trudy Zoologicheskogo Instituta, Akademiya Nauk SSSR, Leningrad 5*, 87–517.

Bernacsek, G.M. & Dineley, D.L. 1977: New acanthodians from the Delorme Formation (Lower Devonian) of N.W.T., Canada. *Palaeontographica A 158*, 1–25.

Burrow, C.J. 2002: Lower Devonian acanthodian faunas and biostratigraphy of south-eastern Australia. *Memoirs of the Association of Australasian Palaeontologists 27*, 75–137.

Burrow, C.J. & Young, G.C. 2002: Diplacanthid acanthodians from the Aztec Siltstone (late Middle Devonian), Antarctica. *Geological Society of Australia, Abstracts 68*, 194.

Campbell, K.S.W. & Barwick, R.E. 1987: Paleozoic lungfishes – a review. *Journal of Morphology, Supplement 1 (1986)*, 93–131.

Dean, B. 1907: Notes on acanthodian sharks. *American Journal of Anatomy 7*, 209–222.

Debenham, F. 1921: The sandstone, etc., of the McMurdo Sound, Terra Nova Bay, and Beardmore glacier regions. *British Antarctic (Terra Nova) Expedition, 1910. Natural History Report (Geology) 1*, 101–119.

Denison, R.H. 1978: Placodermi. *In* Schultze, H.-P. (ed.): *Handbook of Paleoichthyology*, Vol. 2. Gustav Fischer, Stuttgart.

Denison, R.H. 1979: Acanthodii. *In* Schultze, H.-P. (ed.): *Handbook of Paleoichthyology*, Vol. 5. Gustav Fischer, Stuttgart.

Gagnier, P.-Y. 1996: Acanthodii. *In* Schultze, H.P. & Cloutier, R. (eds): *Devonian Fishes and Plants of Miguasha, Quebec, Canada*, 149–164. Verlag Dr Friedrich Pfeil, Munich.

Gagnier, P.-Y., Hanke, G.F. & Wilson, M.V.H. 1999: *Tetanopsyrus lindoei* gen. et sp. nov., an Early Devonian acanthodian from the Northwest Territories, Canada. *Acta Geologica Polonica 49*, 81–96.

Gagnier, P.-Y. & Wilson, M.V.H. 1996: An unusual acanthodian from northern Canada: revision of *Brochoadmones milesi*. *Modern Geology 20*, 235–251.

Gross, W. 1930: Die Fische des Mittleren Old Red Süd-Livlands. *Geologische und Palaeontologische Abhandlungen 18*, 121–156.

Gross, W. 1933a: Die unterdevonischen Fische und Gigantostraken von Overath. *Abhandlungen der Preussischen Geologischen Landesanstalt 145*, 41–77.

Gross, W. 1933b: Die Fische des Baltischen Devons. *Palaeontographica A 79*, 1–74.

Gross, W. 1937: Die Wirbeltiere des rheinischen Devons. Teil II. *Abhandlungen der Preussischen Geologischen Landesanstalt 176*, 1–83.

Gross, W. 1971: Downtonische und Dittonische Acanthodier-Reste des Ostseegebietes. *Palaeontographica A 136*, 1–82.

Gross, W. 1973: Kleinschuppen, Flossenstacheln und Zähne von Fischen aus europäischen und nordamerikanischen Bonebeds des Devons. *Palaeontographica A 142*, 51–155.

Gunn, B.M. & Warren, G. 1962: Geology of Victoria Land between the Mawson and Mulock Glaciers, Antarctica. *Trans-Antarctic Expedition 1955–1958. Scientific Reports, Geology 11*, 1–157.

Hanke, G.F., Davis, S.P. & Wilson, M.V.H. 2001: New species of the acanthodian genus *Tetanopsyrus* from northern Canada, and comments on related taxa. *Journal of Vertebrate Paleontology 21*, 740–753.

Hills, E.S. 1931: The Upper Devonian fishes of Victoria, Australia, and their bearing on the stratigraphy of the state. *Geological Magazine 68*, 206–231.

Janvier, P. 1996: *Early Vertebrates*. Oxford University Press, Oxford.

Lelièvre, H., Janjou, D., Halawani, M., Janvier, P., Muallem, M.S.A., Wynns, R. & Robelin, C. 1994: Nouveaux vertébrés (Placodermes, Acanthodiens, Chondrichthyens et Sarcoptérygiens) de la formation de Jauf (Dévonien inférieur, région de Al Huj, Arabie Saoudite). *Compte-Rendus de l'Academie de Sciences Paris 319, sér. II*, 1247–1254.

Long, J.A. 1983: A new diplacanthoid acanthodian from the Late Devonian of Victoria. *Memoirs of the Association of Australasian Palaeontologists 1*, 51–65.

Long, J.A. 1995: A new groenlandaspidid arthrodire (Pisces: Placodermi) from the Middle Devonian Aztec Siltstone, southern Victoria Land, Antarctica. *Records of the Western Australian Museum 17*, 35–41.

Long, J.A. & Young, G.C. 1988: Acanthothoracid remains from the Early Devonian of New South Wales, including a complete sclerotic capsule and pelvic girdle. *Memoirs of the Association of Australasian Palaeontologists 7*, 65–80.

Long, J.A. & Young, G.C. 1995: New sharks from the Middle–Late Devonian Aztec Siltstone, southern Victoria Land, Antarctica. *Records of the Western Australian Museum 17*, 287–308.

Lyarskaya, L. 1975: New data about acanthodians from the Middle Baltic. *In* Grigyalis, A.A. (ed.): *Fauna and Stratigraphy of Palaeozoic and Mesozoic of Baltic and Byelorussia*, 227–232. Mintis, Vilnius (in Russian).

McKelvey, B.C., Webb, P.N., Gorton, M.P. & Kohn, B.P. 1972: Stratigraphy of the Beacon Supergroup between the Olympus and Boomerang Ranges, Victoria Land. *In* Adie, R.J. (ed.): *Antarctic Geology and Geophysics*, 345–352. Universitets forlaget, Oslo.

McPherson, J.G. 1978: Stratigraphy and sedimentology of the Upper Devonian Aztec Siltstone, southern Victoria Land, Antarctica. *New Zealand Journal of Geology and Geophysics 21*, 667–683.

Miles, R.S. 1966: The acanthodian fishes of the Devonian Plattenkalk of the Paffrath Trough in the Rhineland. *Arkiv für Zoologi 18*, 147–194.

Miles, R.S. 1973a: Articulated acanthodian fishes from the Old Red Sandstone of England, with a review of the structure and evolution of the acanthodian shoulder-girdle. *Bulletin of the British Museum (Natural History), Geology 24*, 113–213.

Miles, R.S. 1973b: Relationships of acanthodians. *Zoological Journal of the Linnean Society of London 53 (Suppl. 1)*, 63–103.

Owen, R. 1846: *Lecture on the Comparative Anatomy and Physiology of the Vertebrate Animals, Delivered at the Royal College of Surgeons of England. I, Fishes.* Longman, Brown, Green & Longmans, London.

Ritchie, A. 1972: Appendix. Devonian fish. *In* McKelvey, B.C., Webb, P.N., Gorton, M.P. & Kohn, B.P. Stratigraphy of the Beacon Supergroup between the Olympus and Boomerang Ranges, Victoria Land. *In* Adie, R.J. (ed.) *Antarctic Geology and Geophysics*, 351–352. Universitets Forlaget, Oslo.

Ritchie, A. 1975: *Groenlandaspis* in Antarctica, Australia and Europe. *Nature 254*, 569–573.

Turner, S. 1997: Sequence of Devonian thelodont scale assemblages in East Gondwana. *Geological Society of America, Special Publication 321*, 295–315.

Turner, S. & Young, G.C. 1992: Thelodont scales from the Middle–Late Devonian Aztec Siltstone, southern Victoria Land, Antarctica. *Antarctic Science 4*, 89–105.

Valiukevicius, J.J. 1983: Microstructure of scales of Middle Devonian acanthodians. *In* Novitskaya, L.I. (ed.): *Problems of Modern Paleoichthyology*, 42–50. Nauka, Moscow (in Russian).

Valiukevicius, J.J. 1998: Acanthodians and zonal stratigraphy of Lower and Middle Devonian in East Baltic and Byelorussia. *Palaeontographica A 248*, 1–53.

Vergoossen, J.M.J. 1999: Siluro-Devonian microfossils of Acanthodii and Chondrichthyes (Pisces) from the Welsh Borderland/south Wales. *Modern Geology 24*, 23–90.

Warren, A., Currie, B.P., Burrow, C. & Turner, S. 2000: A redescription and reinterpretation of *Gyracanthides murrayi* Woodward 1906 (Acanthodii, Gyracanthidae) from the Lower Carboniferous of the Mansfield Basin, Victoria, Australia. *Journal of Vertebrate Paleontology 20*, 225–242.

Watson, D.M.S. 1937: The acanthodian fishes. *Philosophical Transactions of the Royal Society of London, Series B 228*, 49–146.

White, E.I. 1968: Devonian fishes of the Mawson–Mulock area, Victoria Land, Antarctica, Trans-Antarctic Expedition 1955–1958. *Scientific Reports, Geology 16*, 1–26.

Wilson, M.V.H. 1998: New fossil discoveries bearing on homology of paired fins in Gnathostomata. *Journal of Vertebrate Paleontology 18 (Suppl.)*, 87A.

Woodward, A.S. 1891: *Catalogue of the Fossil Fishes in the British Museum (Natural History). Part II.* 567 pp.

Woodward, A.S. 1892: Further contributions to knowledge of the Devonian fishes of Canada. *Geological Magazine 9*, 481–485.

Woodward, A.S. 1906: On a Carboniferous fish fauna from the Mansfield district, Victoria. *Memoirs of the National Museum, Melbourne 1*, 1–32.

Woodward, A.S. 1921: Fish-remains from the Upper Old Red Sandstone of Granite Harbour, Antarctica. *British Antarctic (Terra Nova) Expedition, 1910. Natural History Report (Geology) 1*, 51–62.

Woolfe, K.J., Long, J.A., Bradshaw, M., Harmsen, F. & Kirkbride, M. 1990: Discovery of fish-bearing Aztec Siltstone (Devonian) in the Cook Mountains, Antarctica. *New Zealand Journal of Geology and Geophysics 33*, 511–514.

Young, G.C. 1982: Devonian sharks from south-eastern Australia and Antarctica. *Palaeontology 25*, 817–843.

Young, G.C. 1988: Antiarchs (placoderm fishes) from the Devonian Aztec Siltstone, southern Victoria Land, Antarctica. *Palaeontographica A 202*, 1–125.

Young, G.C. 1989a: The Aztec fish fauna of southern Victoria Land – evolutionary and biogeographic significance. *Geological Survey of London Special Publication 47*, 43–62.

Young, G.C. 1989b: New occurrences of culmacanthid acanthodians (Pisces, Devonian) from Antarctica and southeastern Australia. *Proceedings of the Linnean Society of New South Wales 111*, 12–25.

Young, G.C. 1991: Fossil fishes from Antarctica. *In* Tingey, R.J. (ed.): *The Geology of Antarctica*, 538–567. Oxford University Press, Oxford.

Young, G.C. 1992: Description of the fish spine. *Geological Society of America, Memoir 170*, 273–278.

Yong, G.C. 1993: Middle Palaeozoic macrovcrtebrate biostratigraphy of eastern Gondwana. *In* Long, J.A. (ed.): *Palaeozoic Vertebrate Biostratigraphy and Biogeography*, 208–251, Belhaven Press, London.

Young, G.C. 1996: Devonian (chart 4). *In* Young, G.C. & Laurie, J.R. (eds): *An Australian Phanerozoic Timescale*, 96–109. Oxford University Press, Melbourne.

Young, G.C., Long, J.A. & Ritchie, A. 1992: Crossopterygian fishes from the Devonian of Antarctica: systematics, relationships and biogeographic significance. *Records of the Australian Museum Supplement 14*, 1–77.

Young, G.C., Long, J.A. & Turner, S. 1993: Appendix 1: Faunal lists of eastern Gondwana Devonian macrovertebrate assemblages. *In* Long, J.A. (ed.): *Palaeozoic Vertebrate Biostratigraphy and Biogeography*, 246–251. Belhaven Press, London.

Young, V.T. 1986: Early Devonian fish material from the Horlick Formation, Ohio Range, Antarctica. *Alcheringa 10*, 35–44.

Zidek, J. 1976. Kansas Hamilton Quarry (Upper Pennsylvanian) *Acanthodes*, with remarks on the previously reported North American occurrences of the genus. *University of Kansas Paleontological Contributions 83*, 1–41.

The phylogenetic relationships between actinolepids (Placodermi: Arthrodira) and other arthrodires (phlyctaeniids and brachythoracids)

VINCENT DUPRET

Dupret, V. **2004 06 01**: The phylogenetic relationships between actinolepids (Placodermi: Arthrodira) and other arthrodires (phlyctaeniids and brachythoracids). *Fossils and Strata*, No. 50, pp. 44–55. France. ISSN 0300-9491.

The phylogenetic relationships between actinolepids (Placodermi, Arthrodira) and other arthrodires (Phlyctaenii and Brachythoraci) are re-evaluated in connection with a study in progress on some Podolian and Spitsbergen actinolepid material. The species *Kujdanowiaspis podolica* (Brotzen, 1934) is briefly redescribed, and its characters discussed in relation to a recent cladistic analysis of the actinolepids. An enlarged phylogenetic analysis using a data matrix of 55 characters for 31 taxa reveals that the actinolepids may be a paraphyletic group; their taxonomic and nomenclatural status is briefly discussed. Within the actinolepids, the genus *Kujdanowiaspis* does not have the basal position proposed by previous authors. A proposed new phylogeny shows *Wuttagoonaspis* as the sister group to all other arthrodires, and the Phyllolepida as the sister group to the Phlyctaenioidei (Phlyctaenii plus Brachythoraci).

Key words: Placodermi; Arthrodira; Actinolepida; Phyllolepida; *Kujdanowiaspis*; *Wuttagoonaspis*; phylogenetic analysis.

Vincent Dupret [dupret@mnhn.fr], Département Histoire de la Terre UMR 5143 – USM 0203 CNRS, Laboratoire de Paléontologie, 8 rue Buffon, Muséum National d'Histoire Naturelle, F 75005 Paris, France

Introduction

Within the placoderm order Arthrodira, two complexes have been distinguished: the "dolichothoracids", with a long trunk armour (dermal shoulder girdle) and the "brachythoracids", with a short trunk armour (Denison 1978, 1983, 1984; Gardiner 1984; Goujet 1984a, b, 2001; Goujet & Young 1995; Janvier 1996). The term "Dolichothoraci" was proposed by Stensiö (1944) to replace "Acanthaspida", as both Gross (1937) and Heintz (1937) had shown that the genus *Acanthaspis* was not an arthrodire, and belonged to the order Petalichthyida. The "Dolichothoraci" was a subdivision of the "euarthrodires" (*sensu* Stensiö), members of which possessed "a comparatively long exoskeletal scapular girdle" (Stensiö 1944, footnote).

More recently, the dolichothoracids have been considered not to be a natural group (Goujet 1984a), but corresponding to an evolutionary "grade" covering two major subgroups: the Actinolepidoidei and the Phlyctaenii as defined by Miles (1973). The Actinolepidoidei are the subject of this paper. Goujet (1984a) considered the Phlyctaenii to be more closely related to the Brachythoraci than to the actinolepidoids, and he put the Phlyctaenii and Brachythoraci together within the group Phlyctaenioidei.

My present purpose is to re-evaluate the phylogenetic relationships between the other arthrodires and the

Abbreviations used in figures

ADL, anterior dorsolateral plate; AL, anterolateral plate; AMV, anterior median ventral plate; AV, anteroventral plate; AVL, anterior ventrolateral plate; C, central plate; cc, central sensory line groove; d.end.f, external foramen for the endolymphatic duct; ESC, extrascapular plate; IL, interolateral plate; ioc, infraorbital sensory line groove; lat, lateral scute; lc, main lateral line groove; M, marginal plate; MD, median dorsal plate; mpl, median pit line; N, nuchal plate; occ, occipital cross commissure; PaN, paranuchal plate; PDL, posterior dorsolateral plate; pelv, pelvic girdle; Pi, pineal plate; PL, posterolateral plate; PM, post-marginal plate; PMD, post-median dorsal plate; PMV, posterior median ventral plate; PMV.sc, post-median ventral scute; ppl, posterior pit line; PrO, preorbital plate; PtN, postnasal plate; PtO, postorbital plate; PVL, posterior ventrolateral plate; R, rostral plate; SM, submarginal plate; soc, supraorbital sensory line groove; Sp, spinal plate.

actinolepids. Many authors have attributed a different phylogenetic status to this group. According to some (e.g. Long 1984), actinolepids are a clade, characterised by some special endocranial structures (e.g. supraorbital and basal processes, plus the presence of an endoskeletal internarial process; Goujet 1984a). The sensory lines are deeply sunk in grooves in dermal bones, and the possession of a pair of anteroventral plates on the plastron was considered as unique to this clade (e.g. Miles 1973) until Liu (1991) demonstrated the presence of these bones in the petalichthyid *Eurycaraspis incilis* from China.

These fishes generally have a nasocapsular endocranial region which is very loosely connected to the post-ethmoid ossification (hence their nickname "loose-nose fishes"). However, endocranial material is very rare compared with the number of known actinolepid taxa, and the lack of data means that it is questionable that this group can be considered a clade. Both Miles (1969) and Young & Gorter (1981) argued that the possession of a sliding neck joint (articulation between the skull roof and the thoracic armour) is plesiomorphic when compared with the ginglymoid articulation occurring in the Phlyctaenioidei. The sliding neck joint is also found in phyllolepids. Alternatively, this type of neck joint has been interpreted as a specialisation of actinolepids (Miles 1973; Long 1984). However, the possession of anteroventral plates, considered a primitive character here, would have been lost by all the members of the clade Phlyctaenioidei (Denison 1984).

Most of what has been written on the anatomy of actinolepids refers to Stensiö's works. He was the initiator of a thorough investigation of this group, basing his research on material from Podolia (Ukraine), which is still very incompletely known. This material, including new undescribed specimens, is the foundation for the present study. Podolia is a province situated in southwestern Ukraine. The main river, the Dnister, marks the boundary with Moldavia, and Devonian rocks are exposed all along the Dnister and its tributaries (Fig. 1). The tectonic structure is a slight monocline to the west.

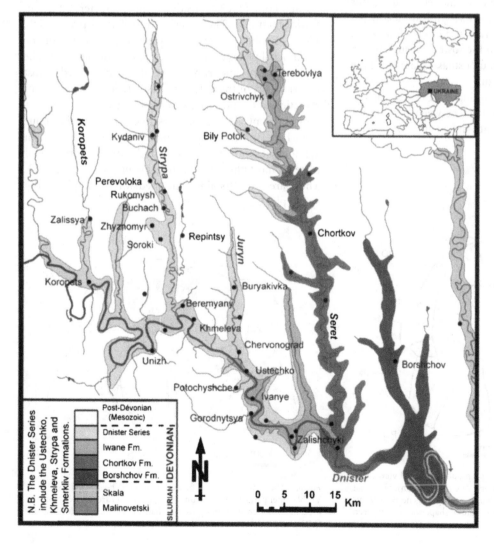

Fig. 1. Devonian deposits of Podolia (Ukraine), redrawn after Blieck (1984). Subdivision of the Dnister Series after Narbutas (1984).

More material was collected in recent field work in Podolia (September 2002).

The Lower Devonian strata in Podolia have yielded many isolated remains of the actinolepid *Kujdanowiaspis* Stensiö, 1942, a genus proposed to include seven species created by Brotzen (1934): *Phlyctaenaspis buczaczensis* (type species), *P. podolica*, *P. rectiformis*, *P. extensa*, *Acanthaspis prominens*, *A. vomeriformis* and *A. angusta*. Stensiö (1942, 1944, 1945) and Denison (1978) included these species in the genus *Kujdanowiaspis*, with the type species *Kujdanowiaspis buczacziensis* (Brotzen, 1934). Our focus here will be limited to the species *Kujdanowiaspis podolica* and *Kujdanowiaspis buczaczensis*. Their detailed description will be the subject of another paper.

Morphology of the genus *Kujdanowiaspis* Stensiö, 1942

The species *Kujdanowiaspis podolica* is the best represented form in the Podolian material, and is briefly described here as a representative model of the genus *Kujdanowiaspis*, by summarising its significant features.

The skull roof (Fig. 2) is characterised by the following:
- no strong cohesion between the rostral capsule, which groups the rostral, pineal and postnasal plates, and the anterior part of the skull roof;
- no contact between the marginal and central plates, which are separated by contact between a posterior process of the postorbital plate and an anterior process of the paranuchal;

- the posterior edge of the nuchal plate is devoid of ornamentation, suggesting the presence of an extrascapular plate;
- the sensory line triple junction between infraorbital and central grooves is located next to the centre of radiation of the post-orbital plate (this is an arthrodire feature);
- the posterior end of the central groove is close to the centre of radiation of the central plate, as well as the ends of the middle and posterior pit lines;
- the infraorbital and main sensory line grooves run along the mesial margin of the marginal plate;
- the external endolymphatic foramen is far in front of the posterior edge of the paranuchal plate, at the geometric centre of the plate (an "actinolepid" feature); it is located next to the posterior pit line and occipital cross commissure, and the centre of radiation of the paranuchal plate (an arthrodire feature);
- the nuchal plate is lanceolate in shape.

The thoracic armour (Figs. 3–5) contains a median dorsal plate (MD, Fig. 3) which is as wide as long, a typical feature for actinolepids within the arthrodires, even though Long (1984) considered this character not useful phylogenetically. The anterior dorsolateral plate (ADL, Figs. 3, 5) shows anterior flanges which slide under the paranuchal plates ("sliding neck joint articulation"). The main sensory line runs from the ornamented, most anterior part of the anterior dorsolateral to the centre of radiation of the posterior dorsolateral plate.

Anteroventral plates are present on the ventral armour (AV, Figs. 3, 4). The spinal plates are quite long for an actinolepid, and can be compared with the long spinal plates of most phlyctaeniids (e.g. *Heintzosteus*, *Arctolepis*). There are spinelets along the medial free margin of the spinal plate.

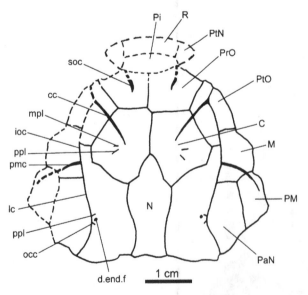

Fig. 2. The skull roof of *Kujdanowiaspis podolica* in dorsal view (after specimen NHM P20773; Natural History Museum, London). Note that this specimen has been excessively compressed by diagenesis, particularly lateral to the main sensory line; its natural shape would have been much more convex.

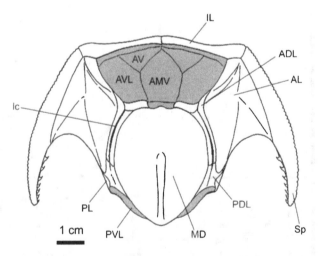

Fig. 3. The thoracic armour of *Kujdanowiaspis podolica*, dorsal view. Visible inner parts of the plastron (ventral armour) are shaded. Note the median dorsal crest in the posterior half of the median dorsal plate (MD) and mesial spinelets on the spinal plate (Sp).

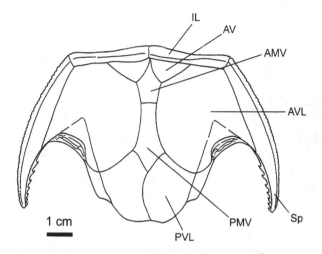

Fig. 4. The thoracic armour of *Kujdanowiaspis podolica*, ventral view.

One specimen of *Kujdanowiaspis* (NASU 25567; State Museum of Natural History, Lviv, Ukraine) shows the presence of four smaller plates (post-median dorsals; PMD, Fig. 5) behind the main median dorsal plate. The second and fourth of these plates show insertion scars supposedly for short spines, but these have not been recovered. The ventral side of this specimen shows another four small plates, with much thinner tubercles (PMV.sc, Fig. 5), and the visceral side of lateral scales are preserved, recalling those described by Goujet (1973) in *Sigaspis lepidophora*.

Phylogenetic analysis of Johnson *et al.* (2000)

Johnson *et al.* (2000) described a new actinolepid (*Aleosteus eganensis*), and compiled a data matrix to present the first cladistic analysis of the actinolepids. Some points to be discussed here concern certain homologies they proposed, their character definitions, and the taxon sample used for the cladistic analysis.

Johnson *et al.* (2000) identified a profundus sensory line on the skull roof of *Aleosteus eganensis*. By comparison with *Bryantolepis*, a form in which a true profundus line is clearly visible (Denison 1958), I consider this line in *Aleosteus eganensis* as a ramification of the infra-orbital sensory line. In *Bryantolepis*, the profundus line is antero-mesially oriented, but in *Aleosteus eganensis* the line on the postorbital plate is oriented anterolaterally, as is normal for the infraorbital groove.

Johnson *et al.* (2000, fig. 1C–E) also illustrated a plate interpreted as an extrascapular, but the specimen seems too strongly arched dorsally (lateral view) for this interpretation. The bone looks more like a terminal postmedian dorsal plate, such as is found in *Kujdanowiaspis podolica* (Fig. 5).

The phylogenetic analysis of Johnson *et al.* (2000) has many characters that are ambiguously coded. For example, those referring to the infraorbital and profundus sensory lines (characters 12, 22–24) are coded only for the derived state; the primitive condition is given as "derived condition not displayed", which does not correspond to a clearly defined state. Characters 17–24 are autapomorphies, and therefore useless for a phylogenetic analysis, but these authors keep them for specific terminal information.

Johnson *et al.* (2000) selected the taxon Petalichthyida as the outgroup, their ingroup comprising only the actinolepids. The Petalichthyida (with Ptyctodontida) are commonly accepted as an outgroup for the Arthrodira (Goujet & Young 1995), but as Arthrodira are composed of actinolepids, phlyctaeniids and brachythoracids, and the ingroup as defined herein is only composed of actinolepids, there is no possibility to test the monophyly of the actinolepids. On the contrary, the cladogram can just reflect a monophyletic status for the actinolepids. This problem has been pointed out by Johnson *et al.* (2000).

New phylogenetic analysis

Using new data from the Podolian material of *Kujdano-wiaspis*, and a re-evaluation of some homologies, I have conducted a new phylogenetic analysis in order to test the monophyly versus paraphyly of actinolepids.

Homology re-evaluations

Among placoderms, phyllolepids are very peculiar organisms because of their dorsoventrally flattened body, and much enlarged "centronuchal plate" on the skull roof. There has been disagreement on the origin of this large unit. It could have resulted from fusion of the nuchal and central plates, or from disappearance of the centrals and expansion of the single nuchal plate. However, this is not an issue for a phylogenetic analysis, which is based on patterns rather than processes. For the phyllolepid taxa, any character specific to the central plates has been coded in the matrix as "not applicable" (NA).

Another problem concerning phyllolepids consists in the primary homologies (*sensu* de Pinna 1991) defined with other arthrodires, i.e. the ring of perinuchal plates. Postorbital, marginal and paranuchal plates are clearly identified (Ritchie 1984; Long 1984). Of the interpretation of the most anterior pairs of plates, Long (1984) and Ritchie (1984) have different views. According to Ritchie, the first pair are post-nasal plates and the second pair preorbital plates. According to Long, the first pair are preorbitals and the second are homologous to the

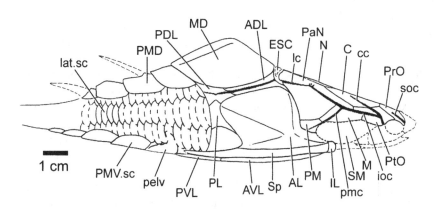

Fig. 5. The skull roof and thoracic armour of *Kujdanowiaspis podolica*, reconstructed in right lateral view. Note the post-median dorsal (PMD) and ventral (PMV.sc) plates, and lateral scales (lat.sc) covering part of the tail.

post-nasals of other arthrodires. I follow Long's assessment (subsequently agreed by A. Ritchie 2002, pers. comm.), with different reasons: generally the two preorbital plates have a mesial junction, and usually contact the central plates, or at least the centronuchal area. The flexure of the supraorbital sensory line on the post-nasal plate is explained by the shift of the plate from its anterior primitive position (as in arthrodires) to a lateral or anterolateral position, and next to the preorbitals in phyllolepids. This shift induced a splitting of the supraorbital sensory line.

Similar problems apply for homologising skull roof bones for the species *Wuttagoonaspis fletcheri* Ritchie, 1973. Like phyllolepids, *Wuttagoonaspis* lacks central plates. But one specimen (AM F54229; Australian Museum, Sydney) shows a suture between the central and nuchal plates, and I consider this specimen as representative of the species and genus, to facilitate the coding of characters. Thus, in contrast to phyllolepids, for *Wuttagoonaspis* characters specific to the central plates have not been coded as "not applicable".

The dermal ornament in both taxa has not been coded, for several reasons. I do not consider the ornament ridges in phyllolepids and wuttagoonaspids as homologous; phyllolepid ornamentation consists of thin and concentric bony ridges, whereas *Wuttagoonaspis* shows both tubercles (which may include semidentine) and ridges, and the ridges are less concentric, showing rather a "fingerprint" pattern.

Matrix construction

The new ingroup used here is composed of the best known actinolepids (17 species), plus phlyctaeniids (five species) and brachythoracids (three species). To complete the spectrum of the Arthrodira, one wuttagoonaspid (*Wuttagoonaspis fletcheri*) and three phyllolepids have been included. The goal was to look at the branching pattern of these groups.

Following Johnson *et al.* (2000), the taxon Petalichthyida is used as the outgroup. However, two petalichthyid species have been scored in the data matrix: *Lunaspis*

broilii and *Eurycaraspis incilis* (Broili 1929; Liu 1991). *Eurycaraspis* was chosen because of the possession of some "actinolepid" features, such as the presence of anteroventral plates. The full listing of 55 characters and their states is given in Appendix 1 (character state numbers do not indicate any a priori choice between the primitive or derived condition). These are scored for 31 taxa in the data matrix shown in Appendix 2.

Data treatment

The phylogenetic analysis was performed using PAUP 3.1.1 (Swofford 1993). The heuristic algorithm was used, because of the number of taxa (data matrix 31 taxa × 55 characters). All characters were unordered and unpolarised a priori, and trees were rooted with the two petalichthyid taxa as an outgroup. Wagner parsimony was used because it accepts both reversions and convergences. The optimisation of missing data was carried out using ACCTRAN (favouring reversions).

The program produced two equal length trees, with length (L) = 137 steps, consistency index (CI) = 0.467 and retention index (RI) = 0.700. The strict consensus from these trees is presented in Fig. 6 (L = 138, CI = 0.464, RI = 0.695). All results discussed below are with reference to this strict consensus tree.

The only polytomy is within brachythoracids (node 28), the monophyly of which is supported. Their sister group is the clade Phlyctaenii (node 24). Both constitute the clade Phlyctaenioidei (node 23). The sister group of this clade is the taxon Phyllolepida. Its position is discussed below. *Wuttagoonaspis fletcheri* appears to be the sister group of other arthrodires (node 3). The actinolepids come out as a paraphyletic group, "basal" among Arthrodira, within which some traditional groupings can still be identified as clades. The family Actinolepididae (node 16), erected by Gross (1940), contains the genus *Actinolepis* Agassiz, 1844, and the species *Bollandaspis woschmidti* Schmidt, 1976. The same grouping emerged in the analysis of Johnson *et al.* (2000), but with a different character history. Other shared characters of

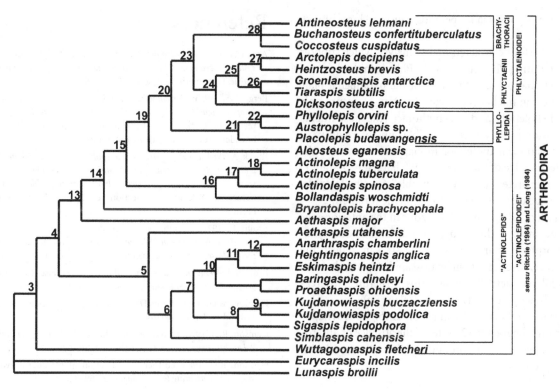

Fig. 6.   The strict consensus tree (length = 138 steps, consistency index = 0.464; retention index = 0.695) resulting from a PAUP analysis producing two equally parsimonious trees (length = 137; consistency index = 0.467; retention index = 0.700). The names of the main groups of arthrodires used in the text are shown on the right.

Actinolepis and *Bollandaspis* are the unpaired preorbital plate and the fused rostral and pineal plates.

The family Kujdanowiaspididae, erected by Berg (1955; "Kujdanowiaspidae") but considered a synonym of Actinolepidae by Obruchev (1964), is defined by node 8. In Fig. 6 this family includes the genus *Kujdanowiaspis* (as defined above) plus the species *Sigaspis lepidophora*. This grouping is supported by a shallow pectoral sinus (long anterior ventrolateral plate spine) and the length/height ratio of the posterior dorsolateral plate > =2 (characters 41, 52; Appendix 1).

The genus *Aethaspis* emerges as highly paraphyletic, with its two species not sharing the same immediate common ancestor. This result is quite surprising, because both species possess a long nuchal plate separating the central plates (character 21, Appendix 1). This state has been treated as a plesiomorphy by PAUP, and it is noted that both Miles (1973) and Young (1980) had already considered the possession of a long nuchal plate as the primitive state. However, the optimisation by PAUP of other characters divides the species between two separate clades. *Aethaspis major* is united with the "upper clade" at node 13, because of postnasal plates connected to the preorbitals (character 9; coded only for *Aethaspis major*, missing data for *Aethaspis utahensis*). *Aethaspis utahensis* is united with the "inner actinolepidoid clade" at node 5, because of the possession of supraorbital and basal

endocranial processes, and no contact between the central and marginal plates (characters 3, 4, 20; Appendix 1). In fact, we do not know about the endocranium for both species, so it appears that the software program optimised the missing data (characters 3, 4) differently for each species. Character 20 is the only one coded differently for the two species, so I am not confident in the implication of Fig. 6 that this genus should be split in two. Indeed, both species, plus *Simblaspis cachensis*, do have a pineal plate but no rostral (D. Goujet 2002, pers. comm.).

In Fig. 6 the Phyllolepida (node 21) are the sister group of the Phlyctaenidoidei (node 20). They are situated at the top of the "actinolepid grade". However, Long (1984, p. 304) considered them as belonging to the Actinolepidoidei, supported by two characters (sliding dermal neck joint; marginal plate generally smaller than postorbital). Long (1984) also interpreted wuttagoonaspids to be close to phyllolepids because of a reduction of the orbits, the small size of marginal plates, and an undifferentiated centronuchal area (but he considered this last character "too unstable for the study of relationships").

Phyllolepids are placed at the top of the "actinolepid sequence" in this analysis because all the characters they share with actinolepids come out as plesiomorphies, and because of very peculiar features (one huge "centronuchal" plate, orbits placed beneath the post-orbital plate, flattened body, lack of post-median ventral plate, etc.)

they form a good clade, perhaps causing "long branch attraction".

There are four characters in this analysis which support node 20 uniting Phyllolepida with Phlyctaenidoidei:

1. an external posterior process of the paranuchal plate behind the nuchal (character 32, state 1) occurs only in Phyllolepida and Brachythoraci (the post-nuchal expansion in Phlyctaenii is visible only internally);
2. the position of the endolymphatic foramen near the posterior edge of the paranuchal (character 33, state 1) was interpreted as "not applicable" for Phyllolepida, which lack an external endolymphatic foramen;
3. the lack of an unornamented zone in the anterior edge of the median dorsal plate ("non-ambiguous synapomorphy"; character 37, state 0) can reflect the presence/absence of an extrascapular plate, but no phyllolepid actually shows this structure, whereas some brachythoracids do (e.g. *Holonema*);
4. anteroventral plates absent ("non-ambiguous synapomorphy"; character 42, state 0): the absence of anteroventral plates is therefore apomorphic (compared with their presence).

Conclusions

In the analysis presented here, "Actinolepidoidei" (with or without *Wuttagoonaspis* and/or phyllolepids) is a paraphyletic group. *Wuttagoonaspis* is not at all close to phyllolepids, but appears here as a sister group to all other arthrodires. The Phyllolepida is the sister group of the Phlyctaenioidei.

The resulting strict consensus cladogram (Fig. 6) is not really congruent with stratigraphy, which can be taken to indicate that the fossil record is far from sufficient. In addition, however, there were many problems encountered with the character coding and the definition of homologies, and in particular an important number of missing data, to be clarified by future research to give a better supported result.

Even if actinolepids are paraphyletic, as suggested by this analysis, I do not recommend the "deletion" of this term, as it remains useful for communication in the study of placoderm phylogenetic relationships.

Acknowledgements

The author is grateful to E. Mark-Kurik (for providing specimens) and D. Goujet for his reviewing of the manuscript and drawings. Photographs were taken by D. Serette and Ph. Loubry (Laboratoire de Paléontologie, MNHN, Paris). Gavin Young and Alex Ritchie are thanked for access to collections, and G. Bosquet made the reconstruction of *Kujdanowiaspis podolica* (http://g.bosquet.free.fr; http://insectus.free.fr).

References

Agassiz, L. 1844: *Monographie des poissons fossiles du Vieux Grès Rouge ou système Dévonien (Old Red Sandstone) des Isles Britanniques et de Russie.* Imprimerie Petitpierre, Neuchâtel et Soleure.

Berg, L.S. 1955: *Sistemariboobraznik i rib, ninie jivooshchikh i iskopaiemikh.* [*Classification of Fishes, Both Recent and Fossil*], 2nd edn revised by E. N. Pavolvsky. Travaux de l'Institut de Zoologie de l'Académie des Sciences de l'URSS (in Russian).

Blieck, A. 1984: *Les Hétérostracés Ptéraspidiformes, Agnathes du Silurien-Dévonien du Continent nord-atlantique et des blocs avoisinants: révision systématique, phylogénie, biostratigraphie, biogéographiem:.* Cahiers de Paléontologie (section vertébrés), Editions du CNRS, Paris.

Broili, F. 1929: Acanthaspiden aus dem rheinischen Unterdevon. *Bayerische Akademie der Wissenschaft Mathematisch-Natwissenschaftliche Klasse Sitzungberichte Abteilungen,* 143–163. Jargh, Munich.

Brotzen, F. 1934: Die silurischen und devonischen Fischvorkommen in Westpodolien II. *Paleobiologica 6(1),* 111–131.

Denison, R.H. 1958: Early Devonian fishes from Utah – Part. III. Arthrodira. *Fieldiana, Geology, 11(9),* 461–551.

Denison, R.H. 1978: *Placodermi. In* Schultze, H.-P. (ed.): *Handbook of Paleoichthyology,* Vol. 5. Gustav Fischer, Stuttgart.

Denison, R.H. 1983: Further consideration of placoderm evolution. *Journal of Vertebrate Paleontology 3(2),* 69–83.

Denison, R.H. 1984: Further consideration of the phylogeny and classification of the order Arthrodira (Pisces: Placodermi). *Journal of Vertebrate Paleontology 4(3),* 396–412.

Dineley, D.L. & Liu, Y.-H. 1984: A new actinolepid arthrodire from the Lower Devonian of Arctic Canada. *Palaeontology 27(4),* 875–888.

Gardiner, B.G. 1984: The relationship of placoderms. *Journal of Vertebrate Paleontology 4(3),* 379–395.

Goujet, D. 1973: *Sigaspis,* un nouvel arthrodire du Dévonien inférieur du Spitsberg (*Sigaspis,* a new arthrodire from the Lower Devonian of Spitsbergen). *Palaeontographica A 143,* 73–88.

Goujet, D. 1984a: *Les poissons placodermes du Spitsberg – Arthrodires Dolichothoraci de la Formation de Wood Bay (Dévonien inférieur).* Cahiers de Paléontologie (section vertébrés), Editions du Centre Nationale de la Recherche Scientifique, Paris.

Goujet, D. 1984b: Placoderm interrelationships: a new interpretation, with a short review of placoderm classifications. *Processes of the Linnean Society of New South Wales 107(3),* 211–243.

Goujet, D. 2001: Placoderms and basal gnathostome apomorphies. *In:* Ahlberg, P.E. (ed.): *Major Events in Early Vertebrate Evolution – Palaeontology, Phylogeny, Genetics and Development,* 209–222. Systematics Association Special Volume Series 61,Taylor & Francis, London.

Goujet, D. & Young, G.C. 1995: Interrelationships of placoderms revisited. *Geobios, Mémoire Spécial 19,* 89–95.

Gross, W. 1937: Die Wierbeltiere des rheinischen Devons. Teil II. *Abhandlungen der Preussischen Geologischen Landesanstalt 176,* 1–83.

Gross, W. 1940: Acanthodier und Placodermen aus den *Heterostius*-Schichten Estlands und Lettlands. *Tartu Ülikooli Geoloogia-Instituut Toimetised 60,* 1–88.

Heintz, A. 1937: Die Dowtonischen und Devonischen Vertebraten von Spitzbergen. VI: *Lunaspis*-Arten aus dem Devon Spitzbergens. *Skrifter om Svalbard og Ishavet 72,* 1–23.

Janvier, P. 1996: *Early Vertebrates.* Oxford Science Publications, Clarendon Press, Oxford.

Johnson, H.G., Elliot, D.K. & Wittke, J.H. 2000: A new actinolepid arthrodire (class Placodermi) from the Lower Devonian Sevy Dolomite, East-Central Nevada. *Zoological Journal of the Linnean Society of London 129*, 241–266.

Liu, Y.-H. 1991: On a new petalichthyid, *Eurycaraspis incilis* gen. et sp. nov., from the Middle Devonian of Zhanyi, Yunnan. *In*: Chang, M.-M., Liu Yu-hai & Zhang Guo-rui (eds): *Early Vertebrates and Related Problems of Evolutionary Biology*, 139–177. Science Press, Beijing.

Long, J.A. 1984: New Phyllolepids from the Victoria and the relationships of the group. *Proceedings of the Linnean Society of New South Wales 107(3)*, 263–308.

Mark-Kurik, E. 1973: *Actinolepis* (Arthrodira) from the Middle Devonian of Estonia. *Palaeontographica 143*, 89–108.

Mark-Kurik, E. 1985: *Actinolepis spinosa* n. sp. (Arthrodira) from the Early Devonian of Latvia. *Journal of Vertebrate Paleontology 5(4)*, 287–292.

Miles, R.S. 1969: Features of placoderm diversification and the evolution of the arthrodire feeding mechanism. *Transactions of the Royal Society of Edinburgh 68(6)*, 123–170.

Miles, R.S. 1973: An actinolepid arthrodire from the Lower Devonian Peel Sound Formation, Prince of Wales Island. *Palaeontographica 143*, 109–118.

Narbutas, V.V. 1984: *Krasnotsvietnaya formatsiya nijnego devona Pribaltiki i Podolii* [*Red Formations of the Lower Devonian of the Baltic States and of Podolia*]. Vilnius, Moscow (in Russian).

Obruchev, D.V. 1964: *Osnovy paleontologii* [*Fundamentals in Paleontology. Agnatha, Pisces*]. Nauka, Moscow (in Russian, translation by Israel Program for Scientific Translations, 1967).

Pinna de, M.C. 1991: Concepts and tests of homology in the cladistics paradigm. *Cladistics 7(4)*, 367–394.

Ritchie, A. 1973: *Wuttagoonaspis* gen. nov., an unusual arthrodire from the Devonian of Western New South Wales, Australia. *Palaeontographica A 143*, 58–72.

Ritchie, A. 1984: A new Placoderm, *Placolepis* gen. nov. (Phyllolepidae), from the Late Devonian of New South Wales, Australia. *Proceedings of the Linnean Society of New South Wales 107(3)*, 321–353.

Schmidt, W. 1976: Der rest eines actinolepididen Placodermen (Pisces) aus der Bohrung Bolland (Emsium, Belgien). *Service Géologique de Belgique 14*, 1–23.

Stensiö, E.A. 1942: On the snout of Arthrodires. *Kungliga Svenska VetenskapsAkademiens Handlingar 20(3)*, 1–32.

Stensiö, E.A. 1944: Contributions to the knowledge of the vertebrate fauna of the Silurian and Devonian of Podolia II – Note on two Arthrodires from the Downtonian of Podolia. *Arkiv för Zoologi 35A(9)*, 1–83.

Stensiö, E.A. 1945: On the heads of certain arthrodires. II. On the cranium and cervical joint of the Dolichothoracids (Acanthaspida). *Kungliga Svenska VetenskapsAkademiens Handlingar 22(1)*, 1–70.

Swofford, D.L. 1993: *PAUP – Phylogenetic Analysis Using Parsimony. 3.1.1*. Distributed by the Illinois Natural History Survey, Champaign, Illinois.

White, E.I. 1969: The deepest vertebrate fossil and other arctolepid fishes. *Biological Journal of the Linnean Society of London 1*, 293–310.

Young, G.C. 1980: A new Early Devonian placoderm from New South Wales, Australia, with a discussion of placoderm phylogeny. *Palaeontographica A 167*, 10–76.

Young, G.C. & Gorter, J.D. 1981: A new fish fauna of Middle Devonian Age from the Taemas/Wee Jasper region of New South Wales. *Bureau of Mineral Resources, Geology and Geophysics, Bulletin 209*, 83–147.

Appendix 1

Character List

The numbers in parentheses (*) refer to characters used by Johnson *et al.* (2000).

Neurocranium

1. Connection between endocranial ethmoid and post-ethmoid components
   0: no connection
   1: connection produced by osseous trabecules or fusion.

No connection is observed in all major groups of arthrodires (Goujet 1984a): in "actinolepids" (e.g. *Kujdanowiaspis*, *Simblaspis*, *Anarthraspis*, *Proaethaspis*, *Heightingtonaspis*), in early phlyctaeniids (e.g. *Arctaspis*, *Gaspeaspis*, *Pageauaspis* (*Quebecaspis*)), and in brachythoracids (only *Buchanosteus*). Connection with osseous trabecules is characterised on the dermal plates by an anterior expansion of the preorbitals (e.g. *Lehmanosteus*), a posterior lengthening of the pineal plate (e.g. *Heintzosteus*) or both (e.g. *Arctolepis*) (Goujet 1984a). Complete fusion is correlated with a loss of the perichondral ossification of the neurocranium (only seen in brachythoracids).

2. Anterior postorbital process
   0: massive
   1: thin.

This character is defined for the dorsal view of the neurocranium (always a thin shape in ventral view). The foramen for the hyomandibular branch of the facial nerve is posterior to this process if it is thin, or is part of the process if it is massive.

3. Supraorbital process
   0: absent
   1: present.
4. Basal process
   0: absent
   1: present.

These processes are present only among actinolepids (Goujet 1984a).

Dermal skull roof

5. Pineal and/or rostral plate(s) separate pre-orbitals (18*)
   0: no separation
   1: separation.

This character refers to the rhinocapsular ossification (complete or not) usually composed of postnasal, pineal and rostral plates. As phyllolepids do not show such a dermal capsule, it was coded "not applicable" (NA; "–" in the data matrix).

6. Rostral and pineal plates are fused
   0: no (separate plates)
   1: yes (single radiation centre).
7. Preorbitals with embayment for attachment of rostropineal or pineal (1*)
   0: no
   1: yes.

The embayment is very shallow for *Baringaspis dineleyi*, *Eskimaspis heintzi*, *Heightingtonaspis anglica*, *Kujdanowiaspis podolica* and *Kujdanowiaspis buczacziensis*, and *Proaethaspis ohioensis*, but there is no doubt that the dermal ethmoid capsule inserted in front of the preorbitals. This shallow embayment is perhaps linked with the loose

connection between the two endocranial components (Goujet 1984a). Nevertheless, the anterior edge of the preorbitals may be superposed with the endocranial optic fissure, or not. I code 1 for these species, and "NA" for phyllolepids (which lack the dermal ethmoid capsule).

8. Rostral and/or pineal fused to the skull roof (2*)
   0: no fusion
   1: fusion.

The pineal plate is fused to the skull roof in some species (*Actinolepis magna*, *Cartieraspis nigra*, *Coccosteus cuspidatus*). I consider this character independent of character 1. As for character 7, the dermal pattern cannot reflect the position of the optic fissure. I coded "NA" for phyllolepids.

9. Postnasals fused to pre-orbitals (4*)
   0: yes
   1: no.

As for *Lunaspis broilii*, I consider that the big plates placed anterolaterally in the skull roof are homologous with arthrodire postnasals. In phyllolepids, the postnasals are the second anterior pair in front of the huge "centronuchal" plate (see text for discussion).

10. Position of orbits on the skull roof
    0: dorsal
    1: lateral.

Only petalichthyids show the dorsal pattern (Denison 1978; Janvier 1996).

11. Preorbital plates (3*)
    0: separate
    1: fused.

Fused preorbitals occur only in *Actinolepis* and *Bollandaspis* (Mark-Kurik 1973, 1985; Schmidt 1976).

12. Sensory lines pattern
    0: canals with superficial pores
    1: grooves.

Arthrodires show a grooved pattern; petalichthyids (with ptyctodontids) have enclosed canals with superficial pores.

13. Supra-orbital lines
    0: separate
    1: meet posteriorly.

The supraorbital grooves meet posteriorly only in *Actinolepis* (Mark-Kurik 1973, 1985). This character is different and not linked to character 2, because *Bollandaspis* shows the separate pattern. Among phyllolepids, only *Austrophyllolepis* shows a junction of the grooves.

14. Infraorbital and main sensory lines run along the margins of the post-orbital and paranuchal plates (15*)
    0: no
    1: yes.

Pattern #1 occurs in *Baringaspis dineleyi*, *Eskimaspis heintzi*, *Heightingtonaspis anglica*, *Kujdanowiaspis*, *Lehmanosteus hyperboreus* and *Sigaspis lepidophora* (White 1969; Goujet 1973, 1984a; Denison 1978; Dineley & Liu 1984; Johnson *et al.* 2000).

15. Differentiated central plates
    0: no (centronuchal plate)
    1: yes.

Central plates are not identified in phyllolepids, but observed in one specimen (AM F54229) of *Wuttagoonaspis* (see text for discussion).

16. Central plates interpenetrate with each other
    0: no
    1: yes.

The derived state is shared by brachythoracids.

17. Pineal and central plates connect
    0: no
    1: yes.
18. Preorbitals indent the anterior margin of centrals (19*)
    0: no
    1: yes.

Johnson *et al.* (2000) only coded 1 for *Eskimaspis heintzi*. I also code 1 for *Coccosteus*, *Groenlandaspis*, *Phlyctaenius* and *Tiaraspis*.

19. Central and preorbital plates (20*)
    0: connected
    1: not connected.

There is no contact between the central and preorbital plates in *Proaethaspis ohioensis*; coded 1 for *Wuttagoonaspis* based on specimen AM F54229.

20. Central and marginal plates (5*)
    0: connected
    1: not connected.

I interpret the central plates and centronuchal area (or centronuchal plates) to be identical for this character. Among phyllolepids, only *Placolepis budawangensis* shows contact between these plates.

21. Nuchal plate separates central plates (6*)
    0: no
    1: yes.

Coded "NA" for phyllolepids, and 1 for *Wuttagoonaspis* after specimen AM F54229.

22. Central plates form part of orbital margin
    0: no
    1: yes.

I consider that central plates in the orbital margin is the primitive state, found only within petalichthyids. Arthrodires share the derived state.

23. Central sensory line reaches the radiation centre of the central plate (11*)
    0: no (stops before radiation centre)
    1: yes.

Only state 1 is explained by Johnson *et al.* (2000). Contrary to these authors, I do not consider this state as derived, but primitive. Indeed, only *Proaethaspis* (Denison 1958, 1978) and *Dicksonosteus* (Goujet 1984a) show a very short central line on the postorbital plate, but not on the central plate. The different skull roof pattern in petalichthyids (lacking a central sensory line) is coded "NA".

24. Posterior pit line on the central and paranuchal plates (10*)
    0: segments connected
    1: segments not connected.
25. Central sensory lines
    0: no central line
    1: on the postorbital plate.

State 0 is encountered in petalichthyids, and state 1 in arthrodires.

26. Postmarginal plate
    0: absent
    1: present.

It is noted that the absence of this plate may be a matter of preservation.

27. Nuchal–central contact suture
    0: nuchal indents centrals
    1: straight (with a large overlap of nuchal onto centrals).

A straight suture (and linked overlap) is encountered only within brachythoracids.

28. Pineal–nuchal plate contact
    0: no contact
    1: contact.
29. Number of paranuchal plates
    1: one pair
    2: two pairs.
30. Occipital cross commissure (8*)
    0: present on paranuchal and nuchal plates
    1: present only on paranuchal plates.
31. Posterolateral edge of paranuchal plates
    0: convex
    1: concave.

The concave condition is encountered only in phlyctaeniids.

32. External dermal process of the paranuchal behind the nuchal
    0: absent
    1: present.
33. External endolymphatic foramen position on paranuchal plates
    0: near geometric centre
    1: near posterior edge.

Phyllolepids are coded "NA" (lack of foramen).

Thoracic armour

34. Structure of dermal craniothoracic articulation
    0: sliding joint (actinolepidoids)
    1: phlyctaeniid ginglymoid-type
    2: brachythoracid ginglymoid-type
    3: spoon-like (petalichthyids).

This character is linked to the position of the para-articular process on the paranuchal plates (lateral to the articular fossa in arthrodires; mesial in petalichthyids).

35. Articular condyles on anterior dorsolateral plates
    0: close together
    1: further apart.

All phlyctaeniids have close condyles; they are widely separated in brachythoracids. Coded "NA" for actinolepids, *Wuttagoonaspis*, phyllolepids and petalichthyids.

36. Ventral keel of median dorsal plate
    0: absent
    1: present.

A ventral keel is typical for brachythoracids.

37. Smooth area on anterior edge of median dorsal (25*)
    0: absent
    1: present.

This smooth area may be for an overlap of an extrascapular plate.

38. Shape of smooth area on median dorsal
    0: simple
    1: double.
39. Extrascapular plate
    0: absent
    1: present.

I consider *Aleosteus eganensis* to lack an extrascapular plate (cf. Johnson *et al.* 2000).

40. Posterolateral plate (38*)
    0: absent
    1: present.
41. Lateral spine on anterior ventrolateral plate (31*)
    0: short
    1: long.

This character reflects the depth of the embayment of the anterior vent-rolateral plate, scored 1 when the extremity extends behind the posterior edge of the anterior ventrolateral.

42. Anteroventral plates (39*)
    0: absent
    1: present.
43. Anterolateral plate contacts anterior ventrolateral behind pectoral fenestra
    0: no contact
    1: contact.
44. Spinelets on the mesial side of the spinal plates (29*)
    0: absent
    1: present.
45. Post-median dorsal plate (26*)
    0: absent
    1: present.
46. Post-median ventral scutes
    0: absent
    1: present.

Quantitative characters

47. (B/L) of preorbitals (17*)
    0: >0.5
    1: <=0.5.
48. Length of central plates (7*)
    0: <45% of skull roof length (without ethmoid capsule)
    1: >=45% of skull roof length (without ethmoid capsule).

49. (L/B) of nuchal plate (14*)
    0: <=1.5
    1: >1.5.
50. (L/B) of median dorsal plate
    0: <1.5
    1: >=1.5.
51. (L/H) of anterior dorsolateral plate (37*)
    0: <1
    1: >=1.

The limit is set at 1, because at 0.5 no actinolepid shows the primitive character state (Johnson *et al.* 2000).

52. (L/H) posterior dorsolateral plate
    0: <2
    1: >=2.
53. The angle between interolateral and spinal plates (32*)
    0: <110°
    1: >=110°.

This character may not be reliable, because of the compaction of many fossils.

54. Length of the spinal/length of the junction anterior ventrolateral plate–spinal plate (28*)
    0: <0.6
    1: >=0.6.
55. (L/B) posterior ventrolateral plate
    0: <1.5
    1: >=1.5.

Appendix 2

character	1	2	3	4	5	6	7	8	9	10	11	12	13	14	15	16	17	18	19	20	21	22	23	24	25	26	27
Antineosteus lehmani	?	?	0	0	0	0	1	1	0	1	0	1	0	0	1	1	0	0	0	0	0	0	1	1	1	1	?
Buchanosteus confertituberculatus	0	0	0	0	0	1	1	1	1	1	0	1	0	0	1	1	0	0	0	0	0	0	1	1	1	1	–
Coccosteus cuspidatus	?	1	0	0	?	0	1	1	0	1	?	1	?	0	1	1	0	1	0	0	0	0	1	1	1	1	1
Arctolepis decipiens	1	1	0	0	1	0	0	1	1	1	0	1	0	0	1	0	1	0	0	0	0	0	1	1	1	1	0
Dicksonosteus arcticus	1	1	0	0	0	0	1	1	1	1	0	1	0	0	1	0	0	0	0	0	0	0	0	1	1	1	0
Groenlandaspis antarcticus	?	?	?	?	1	0	1	1	?	1	0	1	0	0	1	0	1	1	0	0	0	0	1	1	1	1	0
Heintzoteus brevis	1	1	?	?	1	0	1	1	1	1	0	1	0	0	1	0	1	0	0	0	0	0	1	1	1	1	0
Tiaraspis subtilis	?	?	0	0	1	0	1	1	?	1	0	1	0	0	1	0	1	1	0	0	0	0	1	1	1	1	1
Actinolepis magna	1	?	?	?	0	1	1	1	0	1	1	1	1	0	1	0	0	0	0	0	0	0	1	0	1	1	0
Actinolepis spinosa	?	?	?	?	0	1	1	1	?	?	1	1	0	0	1	0	0	0	0	0	0	0	1	0	?	?	0
Actinolepis tuberculata	?	?	?	?	0	1	1	1	0	1	1	1	1	0	1	0	0	0	0	0	0	0	1	0	1	?	0
Aethaspis major	1	?	?	?	0	0	1	0	0	1	0	1	0	0	1	0	0	0	0	0	0	1	0	1	1	1	–
Aethaspis utahensis	1	?	?	?	0	?	1	1	?	1	0	?	0	0	1	0	0	0	0	1	1	0	1	1	1	1	–
Aleosteus eganensis	?	?	?	?	0	?	1	0	?	1	0	1	0	0	1	0	0	0	0	?	0	0	1	1	1	?	0
Anarthraspis sp.	0	?	?	?	0	?	0	0	?	1	0	1	0	0	1	0	0	0	0	1	0	0	1	1	1	1	0
Baringaspis dineleyi	?	?	?	?	0	0	0	0	?	1	0	1	0	1	1	0	0	0	0	1	1	0	1	1	1	1	0
Bollandaspis woschmidti	?	?	?	?	0	0	1	1	0	1	1	1	0	?	?	?	?	0	1	0	?	?	0	1	?	1	?
Bryantolepis brachycephala	?	?	?	?	0	0	1	1	0	1	0	1	0	0	1	0	0	0	0	1	0	0	1	1	1	1	0
Eskimaspis heintzi	?	?	?	?	0	?	0	0	?	1	0	1	0	0	1	0	0	1	0	1	0	0	1	1	1	?	0
Heightingtonaspis anglica	0	0	1	1	0	?	0	0	?	1	0	1	0	0	1	1	0	0	0	0	1	0	0	1	1	1	0
Kujdanowiaspis buczacziensis	0	0	1	1	0	?	0	0	1	1	0	1	0	0	1	0	0	0	0	0	1	0	0	1	1	1	0
Kujdanowiaspis podolica	0	0	1	1	0	?	0	0	1	1	0	1	0	0	1	0	0	0	0	0	1	0	0	1	1	1	0
Proaethaspis ohioensis	0	?	?	?	0	?	0	0	?	1	0	1	0	0	1	0	0	–	1	1	1	1	0	0	1	1	–
Sigaspis lepidophora	?	?	?	?	?	?	?	?	?	?	?	1	?	1	1	0	0	0	0	1	0	1	1	1	1	1	1
Simblaspis cachensis	0	?	?	?	0	?	1	0	?	1	0	1	0	0	1	0	0	0	0	1	0	1	1	1	1	1	1
Phyllolepis orvini	?	?	?	?	–	–	–	–	0	1	0	1	0	0	0	–	–	0	–	1	–	–	1	–	1	0	–
Placolepis budawangensis	?	0	?	?	–	–	–	–	0	1	0	1	0	0	0	–	–	0	–	0	–	–	1	–	1	0	–
Austrophyllolepis sp.	?	0	?	?	–	–	–	–	0	1	0	1	1	0	1	–	–	0	–	1	–	–	1	–	1	0	–
Wuttagoonaspis fletcheri	?	?	?	?	1	0	1	1	?	0	0	1	0	0	0&1	0	0	–	1	0	1	0	1	0	1	1	?
Lunaspis broilii	?	?	0	0	0	0	1	1	1	0&1	0	0	0	0	1	0	0	0	0	0	1	1	–	0	0	0	–
Eurycaraspis incilis	?	?	0	0	1	0	1	1	?	0&1	0	0	1	0	1	0	0	0	1	0	1	0	–	0	0	0	–

character	28	29	30	31	32	33	34	35	36	37	38	39	40	41	42	43	44	45	46	47	48	49	50	51	52	53	54	55
Antineosteus lehmani	0	1	–	1	1	1	2	0	1	1	0	0	?	0	0	?	0	0	?	0	0	1	1	1	1	1	0	?
Buchanosteus confertituberculatus	0	1	0	1	1	1	2	0	1	?	?	?	?	1	?	0	?	?	0	?	1	0	1	1	0	0	?	?	?
Coccosteus cuspidatus	0	1	1	1	1	1	2	0	1	0	–	0	1	0	0	0	0	0	1	0	?	0	?	0	0	1	0	1
Arctolepis decipiens	0	1	0	1	0	1	1	1	0	0	–	0	1	1	0	1	1	0	1	0	1	0&1	1	1	1	1	0	1
Dicksonosteus arcticus	0	1	1	1	0	1	1	1	0	0	–	0	1	1	0	1	1	1	0	1	0	1	1	0	?	0	1
Groenlandaspis antarcticus	0	1	1	1	0	1	1	1	0	0	–	0	1	0	0	1	0	0	1	0	?	1	1	1	0	0	1	1
Heintzoteus brevis	0	1	1	1	0	1	1	1	0	0	0	0	1	1	0	1	0	1	0	1	0	1	1	1	1	1	0	1
Tiaraspis subtilis	0	1	1	1	0	1	1	1	0	0	–	0	1	?	0	1	1	0	?	?	?	?	1	?	?	?	1	1
Actinolepis magna	0	1	1	0	0	0	0	–	0	1	1	?	1	0	1	?	?	0	1	0	0	0	0	1	0	1	?	1
Actinolepis spinosa	0	1	?	?	?	0	0	–	?	?	?	?	?	0	1	?	1	0	1	?	?	?	?	1	?	1	0	0
Actinolepis tuberculata	0	1	0	?	?	0	0	–	?	1	1	?	?	1	1	?	0	0	1	1	?	?	0	1	?	1	0	0
Aethaspis major	0	1	1	0	0	0	0	–	0	1	1	?	1	1	1	0	0	1	1	0	1	1	0	0	0	1	1	?
Aethaspis utahensis	0	1	1	0	0	0	0	–	?	?	?	?	?	1	1	0	0	1	1	0	0	1	?	?	?	1	1	?
Aleosteus eganensis	0	1	1	0	0	0	0	–	0	1	0	?	1	0	1	0	0	1	1	0	0	0	0	1	0	0	0
Anarthraspis sp.	0	1	1	0	0	?	0	–	0	0	–	?	1	1	1	0	0	0	1	0&1	1	1	0	1	0	1	1	?
Baringaspis dineleyi	0	1	1	0	0	0	0	–	0	1	1	?	1	0	0	1	0	0	0	1	0	0	1	0	1	0&1	1	0
Bollandaspis woschmidti	0	?	?	?	?	?	?	–	?	?	?	?	?	?	1	?	?	?	1	0	?	?	?	?	?	?	?	?
Bryantolepis brachycephala	0	1	1	0	0	?	0	–	0	0	–	?	1	1	1	0	0	1	?	0	0	0	1	0	1	1	0
Eskimaspis heintzi	0	1	1	0	0	0	0	–	0	1	1	?	1	1	1	0	0	1	0	1	0&1	0	1	0	0	1	0
Heightingtonaspis anglica	0	1	1	0	0	0	0	–	?	0	0	0	?	0	1	0	0	0	?	0	1	1	?	?	?	0	0	?
Kujdanowiaspis buczacziensis	0	1	1	0	0	0	0	–	0	0	–	0	1	0	1	0	0	1	1	0	0	1	0	1	1	1	?
Kujdanowiaspis podolica	0	1	1	0	0	0	0	–	0	0	–	0	1	0	1	0	0	1	1	0	0	1	0	1	1	1	?
Proaethaspis ohioensis	0	1	1	0	0	0	0	–	0	1	1	?	?	1	1	0	1	0	?	0	1	0	?	0	0	?
Sigaspis lepidophora	0	1	1	0	0	0	0	–	0	1	1	1	1	0	?	?	0	1	1	?	?	1	0	?	?	?	1	0
Simblaspis cachensis	0	1	1	0	0	0	0	–	0	1	1	?	?	?	?	?	?	0	0	0	0	?	?	?	?
Phyllolepis orvini	–	1	1	0	1	–	–	–	0	0	?	0	0	0	0	?	0	0	0	1	–	0	0	1	–	0	0	0
Placolepis budawangensis	–	1	?	0	1	–	0	–	0	0	?	0	0	0	0	?	0	0	0	0	–	0	0	1	–	0	0	0
Austrophyllolepis sp.	–	1	?	0	1	–	0	–	0	0	?	0	0	0	0	?	0	0	0	0	–	0	0	1	–	0	0	0
Wuttagoonaspis fletcheri	0	1	1	0	0	0	0	–	0	?	?	?	?	?	1	?	?	1	?	0	0	1	0	?	?	?	1	?
Lunaspis broilii	1	2	–	0	0	0	3	–	0	0	–	0	0	1	0	?	0	1	1	0	0	1	1	1	0	1	1	0
Eurycaraspis incilis	1	2	1	0	0	?	3	–	0	1	1	1	1	1	1	?	0	0	1	0	0	?	1	1	0	1	0	1

A new genus and two new species of groenlandaspidid arthrodire (Pisces: Placodermi) from the Early–Middle Devonian Mulga Downs Group of western New South Wales, Australia

ALEX RITCHIE

Ritchie, A. 2004 06 01: A new genus and two new species of groenlandaspidid arthrodire (Pisces: Placodermi) from the Early–Middle Devonian Mulga Downs Group of western New South Wales, Australia. *Fossils and Strata*, No. 50, pp. 56–81. Australia. ISSN 0300-9491.

The Merrimerriwa Formation of the Devonian Mulga Downs Group of western New South Wales (NSW), Australia, contains locally rich, dissociated fish remains dominated by an endemic placoderm, *Wuttagoonaspis*, an aberrant arthrodire whose relationships are still uncertain. The *Wuttagoonaspis* fauna of western NSW is now believed to be early Eifelian in age. Associated with *Wuttagoonaspis* in NSW are various actinolepid and phlyctaeniid arthrodires which include a new taxon, *Mulgaspis* gen. nov., described here. *Mulgaspis* shares many characters with Early Devonian phlyctaeniid arthodires such as *Dicksonosteus* and *Arctolepis*, and other characters with the cosmopolitan Late Devonian arthrodiran genus, *Groenlandaspis*. *Mulgaspis* gen. nov., interpreted here as a member of the Groenlandaspididae, provides a morphological link between Early Devonian phlyctaeniid arthrodires and *Groenlandaspis sensu stricto*.

Key words: Placodermi; Arthrodira; Groenlandaspididae; new genus *Mulgaspis*; Devonian; Australia.

A. Ritchie [AlexR@austmus.gov.au], Australian Museum, 6 College Street, Sydney, NSW 2010, Australia

Introduction

Fragmentary Devonian fish remains from the Mulga Downs Group of western New South Wales (NSW) were first reported by Mulholland (1940) and Spence (1958), from sites west of Cobar, NSW. Larger collections of disarticulated fish plates were made by H. O. Fletcher, Australian Museum, and E. O. Rayner, NSW Department of Mines, in 1959 and 1961 from Wuttagoona, Mount Grenfell and Tambua Stations (Fletcher 1964). A preliminary examination of "antiarch, arthrodire and acanthodian remains" from the Mulga Downs fauna

---

Abbreviations used in figures

Placoderm dermal bones: ADL, anterior dorsolateral plate; AL, anterior lateral plate; AMV, anterior median ventral plate; AV, anterior ventral plate; AVL, anterior ventrolateral plate; Ce, central plate; IL, interolateral plate; M, marginal plate; MD, median dorsal plate; Nu, nuchal plate; P, pineal plate; PDL, posterior dorsolateral plate; PL, posterior lateral plate; PM, post-marginal plate; PMV, posterior median ventral plate; PN, post-nasal plate; PNu, paranuchal plate; PrO, pre-orbital plate; Psph, parasphenoid; PtO, post-orbital plate; PVL, posterior ventrolateral plate; R, rostral plate; Sp, spinal plate. Other features: ac, articular condyle; ada, anterodorsal angle (of anterior dorsolateral plate); a.ehy, efferent hyoidean artery; a.ld, laterodorsal aorta; a.pr., anterior process; ant, antorbital process of neurocranium; apo, anterior post-orbital process of neurocranium; app, anterior branch, posterior pitline; art.foss, articular fossa; cc, central sensory canal; d.e., external opening for endolymphatic duct; f.bhy., buccohypophysial foramen; gr.pv., groove for pituitary vein; gr.scc, groove for semicircular canal; hyp.v., groove for hypophysial vein; in.r, internasal ridge; lc, lateral canal; lac, "accessory twig" sensory line (on anterior dorsolateral plate); mpl, middle pitline; n.cap, nasal capsule; n.f., nasal fenestra; oa.ADL, overlap area with anterior dorsolateral plate; oa.MD, overlap area with median dorsal plate; oa.PDL, overlap area with posterior dorsolateral plate; oa.PL, overlap area with posterior lateral plate; occ, occipital cross-commissure pitline; or, ornament of ascending lamina of interolateral plate; orb, orbit; pap, occipital para-articular process; pda, posterodorsal angle (of posterior dorsolateral plate); pbla, post-branchial lamina; pmc, post-marginal sensory canal; poa, post-obstantic area; p.p., pineal pit; ppl, posterior pitline; ppp, posterior branch, posterior pitline; ppo, posterior post-orbital process of neurocranium; scc, scapulocoracoid; spf, subpituitary fossa; VII hm, facial nerve, hyomandibular branch.

suggested that the fauna was Late Devonian (Famennian) in age. Subsequent collections from many sites have failed to confirm the presence of antiarchs in the fauna, and the identification of some of the placoderm plates as *Phyllolepis* also proved to be incorrect.

Rade (1964) reported a similar fauna from Mount Jack, north of Wilcannia, NSW, also based on fragmentary material. The placoderms *Holonema*, *Groenlandaspis*, and *Phyllolepis*, and the acanthodian *Striacanthus*, were again assessed as supporting a Late Devonian (Famennian) age. The identifications of *Holonema* and *Phyllolepis* proved incorrect, and the material attributed by Rade to *Phyllolepis* was later recognised as belonging to a rather unusual new placoderm, *Wuttagoonaspis fletcheri* Ritchie, 1973. This is the most abundant and widespread placoderm in the Mulga Downs fauna. The affinities of *Wuttagoonaspis* have since been the subject of considerable dispute (Miles & Young 1977; Young 1980, 1981; Long 1984; Goujet & Young 1995; Janvier 1996; Young & Goujet 2003).

Associated with *Wuttagoonaspis* at Tambua, Mount Grenfell, and Mount Jack, were small isolated trunk shields attributable to phlyctaeniid euarthrodires, and similar to taxa (e.g. *Huginaspis*) known from the Early Devonian of Spitsbergen (Ritchie 1969, fig. 4). Isolated phlyctaeniid headshields of approximately the same size, recovered from the same sites (Ritchie 1969, fig. 3), were initially assumed, incorrectly, to belong to these trunk shields. Later visits to the same and other sites recovered numerous dissociated trunk shield elements and many more headshields which clearly belonged to the same taxon. These represent a new genus and species of groenlandaspidid, described here as *Mulgaspis evansorum* gen. et sp. nov. Material of a second, but much larger groenlandaspidid, from other localities in the Mulga Downs Group, is clearly related to *Mulgaspis evansorum*, and represents a second species described below as *Mulgaspis altus* sp. nov. Nothing in the *Wuttagoonaspis* fauna of the lower part of the Mulga Downs Group supports a Late Devonian age, and it is now considered to be late Early to early Middle Devonian, i.e. Emsian–Eifelian in age.

The *Wuttagoonaspis* fauna is widespread in Devonian strata across the Australian continent, and Young & Goujet (2003) have recently described similar fish faunas from the Georgina and Amadeus Basins of central Australia, which include a new species of *Wuttagoonaspis*, and distinctive new arthrodire taxa including groenlandaspid remains, some of which may be referable to the material from the Mulga Downs Group described in this paper. The discovery and recognition of many new species of *Groenlandaspis*, and several new genera of groenlandaspidids from Antarctica (Ritchie 1975; Long 1995, *Boomeraspis*), Australia (Ritchie 1974; Young & Goujet 2003, *Mithakaspis*), England, Ireland, Byelorussia (Ritchie 1974), Iran (Janvier & Ritchie 1977), USA

(Daeschler *et al.* 2003, *Turrisaspis*), South Africa (Long *et al.* 1997, *Africanaspis*), and now *Mulgaspis* gen. nov. described below, confirms that the family Groenlandaspididae not only had a very long history, but also a cosmopolitan distribution in the Devonian.

The material described in this paper comes from six localities, one in the Mount Jack area, 60 km north-northeast of the Darling River town of Wilcannia, and the rest from the Tambua-Wuttagoona area of the Dunlops Range, about 60 km west and northwest of the mining town of Cobar. The second area is some 170 km southeast of Mount Jack. A map of the main fossil fish localities was given in Ritchie (1973, fig. 1), and the general Devonian geology of the Darling Basin was summarised by Bembrick (1997, fig. 1).

Throughout this paper these six fossil localities are numbered as follows: (1) Mount Jack Station, north of Wilcannia; (2) Tambua Station; (3) Mount Grenfell Station; (4) Wuttagoona Station; (5) Tambua Station (lower horizon); (6) Mount Grenfell Station (lower horizon). All material described below is housed in the Australian Museum, Sydney (prefix AMF. or F.).

Stratigraphy and age of the *Wuttagoonaspis* fauna in NSW

Glen (1979, 1982a, b, 1987) mapped the Devonian sections in the Buckambool and Wrightville areas, south of Cobar, and subdivided the local Devonian succession (Table 1). Glen's formation names have been adopted by geologists mapping in the Tambua, Mount Grenfell and Wuttagoona areas of the Dunlops Range, but much of this work remains unpublished.

Evidence of local angular unconformity, disconformity, and paraconformity between the Winduck Group and the overlying Mulga Downs Group indicates "lithification, stillstand, uplift, erosion and localised folding in response to block faulting before deposition of the Mulga Downs Group" (Glen 1982a, p. 135).

Glen (1982a, p. 131) interpreted the palaeoenvironment of the Merrimerriwa Formation as fluvial, manifested by "point-bar sequences characteristic of

Table I. Subdivision of the local Devonian succession in the Buckambool and Wrightville areas, south of Cobar (after Glen 1979, 1982a, b, 1987).

Mulga Downs Group	Crowl Creek Formation
	Bundycoola Formation
	Bulgoo Sandstone
	Merrimerriwa Formation
	Meadows Tank Formation
Winduck Group	Gundaroo Sandstone
	Sawmill Tank Sandstone
	Buckambool Sandstone

high-sinuosity (meandering) stream systems. Granule conglomerates at the base of some cycles represent channel lags. Cross beds (dominantly trough-shaped) represent migration of dune bed-forms; their upward decrease in set thickness reflects decrease in current velocity. Siltstones in the upper part of each cycle represent vertical accretion, overbank deposits." The fish remains occur mainly at the base of these upward-fining alluvial cycles. In cosets of cross bed sets the fossil fish remains are most abundant at the base of the cosets, and also sporadically present (but less common) within the foreset laminae of the individual cross bed sets. The fauna is dominated by *Wuttagoonaspis*, with other taxa usually present in minor quantities.

Apart from rare, and mostly unidentifiable, placoderm plate impressions in the Gundaroo Sandstone of the Winduck Group, almost all of the well-preserved Devonian fish remains come from the lower part of the Merrimerriwa Formation of the Mulga Downs Group which consists of a cyclic alternation of sandstone and siltstone. In the absence of other biostratigraphically useful fossils, the fish remains from the lower Mulga Downs Group, and in particular from the underlying Winduck Group, have been misinterpreted and misreported in previous publications concerning the age of these rocks.

Glen (1982a, pp. 131, 133, pers. comm.) incorrectly reported that I had identified *Wuttagoonaspis* plates in the Gundaroo Sandstone at the top of the Winduck Group. In fact, because the few isolated fish plate impressions from the Gundaroo Sandstone are poorly preserved, I had declined to identify them more specifically than "arthrodire indet.". Glen was incorrect in citing my earlier suggestion (Ritchie 1973, p. 70), that *Wuttagoonaspis* was probably of Emsian–Eifelian age, as evidence that the top part of the Gundaroo Sandstone must be of "Early–Middle Devonian" age. A preliminary report on contemporary discoveries of new Devonian fossil fish from western NSW (Ritchie 1969) refuted earlier claims of a Late Devonian (Famennian) age for the lower part of the Mulga Downs Group. In that article it was mentioned that the small phlyctaeniid arthrodires found in the Mulga Downs Group in the Dunlops Range and at Mount Jack show the closest resemblance to phlyctaeniids from Spitsbergen; namely *Huginaspis*, *Heterogaspis* and *Arctolepis*, "which occur in the lower part of the Middle Devonian (Eifelian)" (Ritchie 1969, p. 223; see also Neef *et al.* 1996).

I was subsequently cited by Glen (1992, p. 252) as follows: "A Late Early Devonian age for the lower part of the Mulga Downs Group is based on the association of fossil fish and acritarchs (Ritchie 1969)". In fact acritarchs were nowhere mentioned in my article, and the age suggested was Middle (not Early) Devonian.

Evidence for an Eifelian age for the Mulga Downs *Wuttagoonaspis* fauna has recently emerged from another source. Dr Patrick Conaghan (Macquarie University, Sydney), has studied the Devonian succession in the stratigraphic test bore DM Mossgiel DDH-1 in the Hillston Trough, an infra-sub-basin of the Darling Basin located southwest of the southern end of the Cobar Basin and the northern end of the Mount Hope Trough. The Devonian succession at Mossgiel rests non-conformably on Silurian granite and its top is truncated by the Tertiary sub-Murray Basin erosion surface. According to Conaghan (written communication, March 2003), the Devonian section in DM Mossgiel DDH-1 commences in the latest Pridoli or very earliest Devonian, extends through the whole of the Lower Devonian, and terminates within the Eifelian (= Early Middle Devonian). The age of the Mossgiel succession is constrained by K-Ar radiometric ages on air-fall tuffs throughout the section, by sequence-stratigraphic analysis of the section in comparison with that of the Cobar Basin (=Cobar Supergroup) in relationship to the Euramerican and Australian eustatic curves, and by conodont biostratigraphic ages on limestones within the Cobar Supergroup. The Mossgiel succession is interpreted as marine-turbiditic except for a short interval in the middle, which is terrestrial in the base and possibly paralic in the top. His analysis suggests that the top of the Mossgiel section probably terminates within T-R Cycle Ie of the Euramerican eustatic curve (= late Eifelian).

Conaghan's sequence-stratigraphic analysis of the Cobar Supergroup suggests: (1) that the likely age of the top of the Gundaroo Sandstone, the topmost formation of the Winduck Group, is late Emsian (= *serotinus* Zone); (2) that the base of the Meadows Tank Formation (=lowermost formation of the Mulga Downs Group) is probably latest Emsian (= *patulus* Zone), manifesting the commencement of T-R Cycle Ic on the Euramerican eustatic curve; and (3) that the overlying Merrimerriwa Formation in which the abundant *Wuttagoonaspis* fish remains occur is early–mid-Eifelian, the stratigraphic interval containing the abundant *Wuttagoonaspis* and associated *Lingula* fossils probably manifesting the basal *australis* Zone transgression that commences T-R Cycle Id on the Euramerican eustatic scale. The Merrimerriwa Formation is interpreted as slightly older than the top of the marine-turbiditic succession at Mossgiel, and the brackish-marine incursion recorded within the Merrimeriwa Formation by the presence of *Lingula* and the *Wuttagoonaspis* fauna entered the Winduck Shelf region and beyond from the Hillston Trough. In Conaghan's palaeogeographic reconstruction the Hillston Trough formed a continuous marine seaway with the Melbourne Trough through the Early and Middle Devonian until the time of the Tabberabberan event in the Givetian.

In their recent review of Devonian microvertebrate faunas and their importance for marine–non-marine correlation in East Gondwana, Young & Turner (2000, pp. 461–464) discussed the widespread distribution in eastern Australia of macrovertebrate faunas (MAV2) containing the endemic genus *Wuttagoonaspis*. In the Darling Basin (their fig. 3, col. D; MAV2), as well as in the Amadeus and Georgina Basins (fig. 4, cols C, D; MAV2), the suggested age is Emsian. It is likely, however, that *Wuttagoonaspis* had quite a long range, and the evidence cited above suggests that in the Darling Basin the main occurrences of the *Wuttagoonaspis* fauna are early Eifelian. However, the association of *Wuttagoonaspis* at Mount Jack with abundant thelodont scales referred to *Turinia australiensis* (Turner *et al.* 1981, fig.16) presents a problem, as the range of *Turinia australiensis* is given as somewhat older (*pesavis-?dehiscens* CZs), but may well prove to be more extensive (Young & Turner 2000, p. 461, fig. 2).

At the six numbered localities from which the new genus *Mulgaspis* is described below, the sediments containing *Mulgaspis evansorum* sp. nov. at localities 1–3 (Mount Jack and Dunlops Range areas), are fine-grained quartzites. At localities 4–6 (Dunlops Range area) the sediments containing *Mulgaspis altus* sp. nov. are coarser grained, lie further east in a westerly dipping sequence, and probably represent a lower horizon in the Merrimerriwa Formation.

Systematic palaeontology

Class Placodermi McCoy, 1848

Order Arthrodira Woodward, 1891

Suborder Phlyctaenioidei Miles, 1973

Family Groenlandaspididae Obruchev, 1964

Diagnosis. – Phlyctaeniids in which the pineal plates separate the pre-orbitals; the central sensory lines pass back onto the central plates; the dermal neck joint is situated close to the midline; the anterior and posterior dorsolateral plates meet in the midline under the median dorsal plate; and the median dorsal is elevated as a long, narrow crest or spine. The lateral line sensory groove has a sharp dorsal inflection on the posterior dorsolateral plate.

Remarks. – Young & Goujet (2003, p. 48) noted that many of the characters previously used to define the family Groenlandaspididae (Ritchie 1975, p. 571) are features common to all phlyctaenioid arthrodires. They proposed a shorter, simpler diagnosis for the family, retaining only those characters separating *Groenlandaspis* and a few

closely related taxa from other phlyctaenioids. Their revised version is slightly modified here, to take account of the identification of a separate post-nasal plate in *Mulgaspis* gen. nov. The following genera are included in the family: *Africanaspis, Boomeraspis, Groenlandaspis, Mulgaspis* nov., *Mithakaspis, Tiaraspis,* and *Turrisaspis.*

Genus *Mulgaspis* nov.

Type species. – *Mulgaspis evansorum* sp. nov.

Etymology. – After the Mulga Downs Group, and "*aspis*", Greek for "shield".

Diagnosis. – Groenlandaspidid with headshield slightly wider than long; posterior margin convex and angular. Rostral plate very short and wide, forming most of the anterior margin, with separate post-nasal plates; pineal plate large, tapering anteriorly. Median dorsal plate very narrow and high, with anterior and posterior margins of about equal length. Dorsolateral plates meet in midline, but with anterodorsal margin of anterior dorsolateral and posterodorsal margin of posterior dorsolateral overlapping the median dorsal. Posterior dorsolateral overlaps anterior dorsolateral ventral to lateral line sensory groove, which crosses from the anterior dorsolateral slightly ventral to midpoint on their common suture. Posterior section of lateral line on posterior dorsolateral directed horizontally, not posteroventrally. Post-branchial lamina of interolateral plate ornamented with six to eight denticulate ridges subparallel to the dorsal margin; ventrolateral part of lamina lacks ridges, and is ornamented with fine tubercles. Posterior ventrolateral plates much shorter than anterior ventrolaterals. Anterior median ventral short, subpentagonal, tapering posteriorly, and separated from interolaterals by small, triangular anterior ventral plates which meet mesially. Posterior median ventral longer than anterior median ventral, tapering anteriorly, and widest posteriorly.

Mulgaspis evansorum sp. nov.
Figs. 1–7

Synonymy. –

1969 "headshield of a small arctolepid arthrodire" – Ritchie, p. 222, fig. 3
1975 "new genus from the Mulga Downs Group" – Ritchie, pp. 570–572, fig. 3f
1987 "new genus *Mulgaspis*" (*nomen nudum*) – Ritchie, p. 253 (*In* Glen 1987)
1993 "groenlandaspid nov." – Young *et al.*, p. 247
1996 "new genus of groenlandaspid arthrodire" – Neef *et al.*, p. 21, fig. 3B 1–3

1995 "new undescribed form from the Mulga Downs Group" – Long, p. 39

2002 "new genus of groenlandaspidid arthrodire" – Ritchie, p. 137

2003 "groenlandaspid gen. et sp. nov." (*in pars*) – Young & Goujet, pp. 49–55, fig. 26B

Etymology. – The species is named after Ken and Anne Evans and family, of Tambua Station, near Cobar, in recognition of their generous support to Australian Museum field parties and to international visitors for over 30 years.

Holotype. – AMF.61371, headshield (Fig. 1B, C).

Referred material. – Headshields: AMF.53598, 53666, 54164, 54163, 54643, 61287, 61304, 61348, 61371, 61373, 61621, 64827, 64830, 64836. Dorsolateral trunk plates: AMF.53574 (posterior dorsolateral plate), 53975 (median dorsal, posterior lateral plates), 54152 (median dorsal, anterior dorsolateral, posterior dorsolateral plates), 54162 (anterior lateral plate), 54205 (posterior dorsolateral plate), 61285B (median dorsal, anterior dorsolateral, posterior dorsolateral plates), 61365 (median dorsal, anterior dorsolateral, posterior dorsolateral plates), 61369A (median dorsal, anterior dorsolateral, posterior dorsolateral plates), 61619 (median dorsal plate), 65089 (anterior lateral plate). Ventral trunk plates: AMF.54695 (anterior ventrolateral, anterior ventral, spinal, interolateral plates), 56253 (anterior ventrolateral plate), 56257 (posterior ventrolateral plate), 61282 (anterior ventrolateral, posterior ventrolateral, anterior ventral, spinal, interolateral plates), 61294A (anterior ventrolateral, anterior ventral, spinal, interolateral plates), 61379D (anterior median ventral plate), 61379G (posterior median ventral plate), 61618E (posterior ventrolateral plate), 64828 (interolateral plate), 64830 (anterior ventrolateral, posterior ventrolateral, anterior ventral, spinal, interolateral plates), 65090 (anterior ventrolateral, posterior ventrolateral, spinal, interolateral plates).

Diagnosis. – As for genus, with the following additions. Groenlandaspidid with combined head and trunk shield up to 60 mm long. Headshield up to 40 mm long. Median dorsal ridge with straight anterodorsal and concave posterodorsal margins. Anterior lateral plate longer than high (height/length index 68–73). Posterior ventrolateral much shorter than anterior ventrolateral (length index of anterior/posterior ventrolateral plates about 200). Free pectoral spine of spinal plate of variable length, but usually shorter than fixed portion (25–30%).

Remarks. – Although the collecting sites lie over 170 km apart, they have yielded very similar faunas, in both areas dominated by *Wuttagoonaspis fletcheri* Ritchie, 1973. There is no significant difference between the *Mulgaspis*

samples from the various sites, and they are considered here to represent a single small species, *Mulgaspis evansorum.*

Localities. –
(1) Type locality (Mount Jack Station, 60 km north-northeast of Wilcannia, western NSW, on the crest of a low ridge 2–3 km east of Mount Jack homestead). AMF.54152 (median dorsal, anterior dorsolateral, posterior dorsolateral plates), F.54162 (anterior lateral plate), F.54163, F.54164, F.54218 (nasal capsule), F.56253 (anterior ventrolateral plate), F.56257 (posterior ventrolateral plate), F.61282 (anterior ventrolateral, posterior ventrolateral, anterior ventral, spinal, interolateral plates), F.61285B (median dorsal, anterior dorsolateral, posterior dorsolateral plates), F.61287, F.61294A (anterior ventrolateral, anterior ventral, spinal, interolateral plates), F.61348, F.61365 (median dorsal, anterior dorsolateral, posterior dorsolateral plates), F.61366, F.61369A (median dorsal, anterior dorsolateral, posterior dorsolateral plates), F.61618E (posterior ventrolateral plate), F.61619 (median dorsal plate), F.61621, F.63171, F.64827, F.64828 (interolateral plate), F.64830 (anterior ventrolateral, posterior ventrolateral, anterior ventral, spinal, interolateral plates), F.64830, F.65089 (anterior lateral plate); F.65090 (anterior ventrolateral, posterior ventrolateral, spinal, interolateral plates).

(2) Tambua Station, 70 km west-northwest of Cobar, western NSW, from a low north–south ridge, 3–4 km north of Tambua homestead. AMF.53574 (posterior dorsolateral plate), F.53598, F.53666, F.54205 (posterior dorsolateral plate), F. 54502, F. 54695 (anterior ventrolateral, anterior ventral, spinal, interolateral plates), F.64836.

(3) Mount Grenfell Station, 70 km northwest of Cobar, western NSW, around Bald Hill Tank, 2 km east of Mount Grenfell homestead. AMF.53975 (median dorsal, posterior lateral plates), F.54643 headshield), F.61379D (anterior median ventral plate), F.61379G (posterior median ventral plate).

Horizon and age. – Merrimerriwa Formation, Mulga Downs Group, early Eifelian.

Description. – All material occurs as natural moulds in quartzite. Sometimes there are poorly preserved remains of the original bone, which have been removed mechanically or with acid to clean the impressions for latex casting and study.

Headshield
Relatively complete, uncrushed headshields of *Mulgaspis evansorum* sp. nov. are moderately common, and many

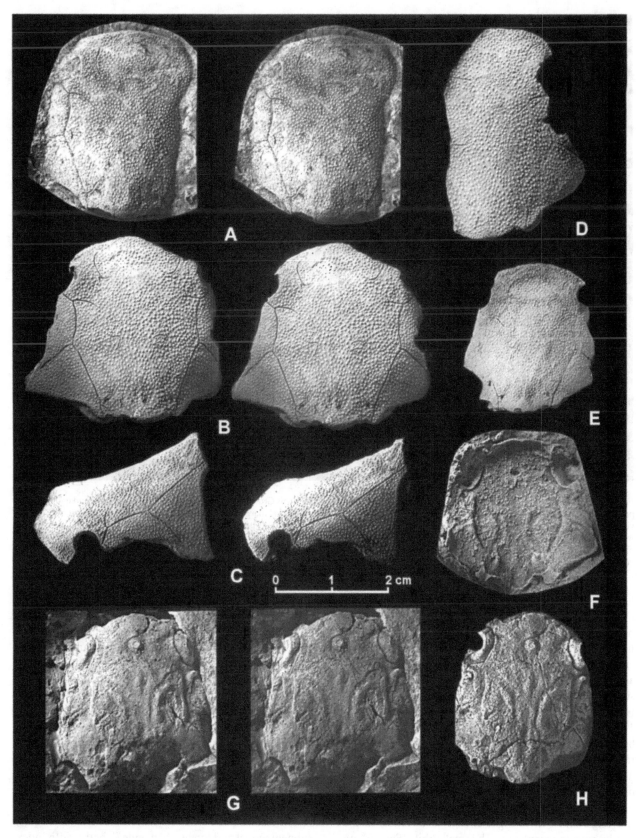

Fig. 1.   Mulgaspis evansorum gen. et sp. nov., Merrimerriwa Formation, Mulga Downs Group. A, B: Headshields, dorsal view, stereo-pairs, AMF.61621 (A), AMF.61371 (B, holotype). C: Headshield in lateral view, stereo-pair (holotype, AMF.61371). D: Incomplete headshield, dorsal view (AMF.61287). E: Headshield, dorsal view (AMF.54643). F: Headshield, visceral view (AMF.53666). G, H: Dorsal view of endocranial moulds of headshield, AMF.64836 (G, stereo-pair), AMF.53598 (H).

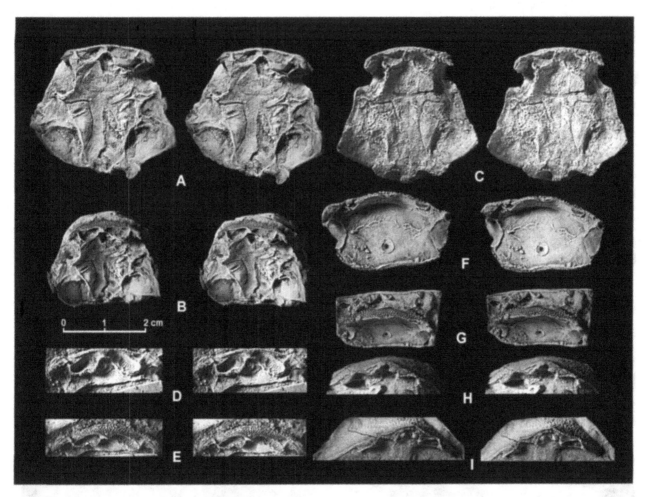

Fig. 2. *Mulgaspis evansorum* gen. et sp. nov., Merrimerriwa Formation, Mulga Downs Group. A, B: Headshield with nasal capsules and endocranium preserved, stereo-pairs, AMF.64830 (A, ventral view; B, inclined obliquely to show rostral margin). C: Headshield with endocranium preserved, ventral view, stereo-pair (AMF.54164). D, E: Anterior portion of incomplete headshield with well-preserved nasal capsules and rostral margin, stereo-pairs, AMF.54163 (D, ventral view; E, oblique anteroventral view). F, G: Anterior portion of incomplete headshields lacking nasal capsules, and showing pineal pit and radiating plexus of blood vessels underlying rostral and pre-orbital plates, stereo-pairs, AMF.64827 (F, ventral view; G, oblique anteroventral view). H, I: Headshields in posterior view showing articular fossae on posterior margin, stereo-pairs (H, AMF.64827; I, AMF.61304).

have now been recovered from the three localities. The original bone, if present, is poorly preserved. Most headshields occur as natural moulds in quartzite, showing the dermal, visceral and, more rarely, the ventral endocranial surfaces. The cranial plates were firmly fused and usually break across, rather than along, the cranial sutures that are only occasionally faintly preserved. Headshields of *Mulgaspis evansorum* range from 28 to 39 mm in length, and from 29 to 41 mm in width. The headshield is marginally wider than long (length/width ratio 86–94), is widest at the posterolateral angle, and narrows anteriorly (Figs. 1, 4). The gently convex rostral margin is about two-thirds the maximum width. The posterior margin is convex, bluntly pointed at the posterior angle and with a slight change of direction at the craniothoracic articulation. The small orbits (Fig. 4E) are situated anterolaterally, and deeply notch the cranial roof. They are directed laterally, and are more visible in lateral than in dorsal view.

The plate pattern is basically like that of several phlyctaeniid arthrodiran genera, but sufficiently different to justify the erection of a new genus. Only the plates of the cranial roof have been recovered, and suborbital and cheek plates have not yet been located or identified in the masses of broken and dissociated arthrodiran plates in the Mulga Downs faunal assemblages. Most of the cranial plate margins are straight or slightly curved, similar to *Arctolepis*, indicating that the overlap between individual plates was minimal, and that *Mulgaspis* had not developed more extensive and complex interplate overlap relationships like those in *Groenlandaspis*. Individual plates are described with reference to the suture pattern shown in Fig. 4E.

The nuchal plate is short (37–41% of cranial length), subpentagonal and longer than wide (length/width index 174–185). It is bluntly rounded posteriorly, acutely pointed anteriorly where it partly separates the posterior parts of the centrals, and the lateral margins are subparallel (Figs. 1A, B, 4E). Posteriorly the nuchal plate rises rather sharply to a bluntly pointed crest (Fig. 1C).

The pineal plate is large, pentagonal, longer than wide (length/width index 118–135) and forms about 30% of the cranial length (Figs. 1A, 4E). It is single, undivided and completely separates the pre-orbitals. It narrows anteriorly (unlike in *Arctolepis* where the pineal plate widens anteriorly) and is widest at the junction with the pre-orbital and central plates. It is bluntly pointed posteriorly where it partly separates the central plates. *Mulgaspis* thus differs from many other actinolepids and phlyctaeniids in which the pineal is either greatly reduced or even fused to the rostral. On the visceral surface of the headshield (where no plate boundaries can be distinguished), a deep pineal pit is centrally situated under the pineal plate (Figs. 1F–H, 2F, G, 3A, 4B).

Daeschler *et al.* (2003) described and figured a new genus of groenlandaspidid, *Turrisaspis elektor*, in which the pineal plate is divided into separate anterior and posterior elements, with the pineal pit situated under the anterior pineal. A similar transverse division into anterior and posterior pineal plates has also been observed in *Groenlandaspis* spp. from Antarctica (*G. antarctica*), Ireland (*G. disjectus*) and in new species from Australia (Mount Howitt, Victoria, and Canowindra, NSW), currently under study by the author. Interestingly, in *Tiaraspis* (Schultze 1984, figs. 1–3) the pineal is depicted as a small plate bordered anteriorly by a large opening, or fontanelle, raising the possibility that it was also divided into anterior and posterior elements and that the anterior one has been lost or perhaps just displaced.

The rostral plate (Fig. 4E) in *Mulgaspis evansorum* is short and wide, forming about 77% of the anterior margin. It consists of two laminae which meet at right angles or even less. The dorsal lamina is flat, or slightly concave, with a low, transverse crescentic ridge developed

Fig. 3. *Mulgaspis evansorum* gen. et sp. nov., Merrimerriwa Formation, Mulga Downs Group. A: AMF.54164, anterior part of headshield (ventral view) showing detail of nasal capsules, pineal pit, and interpreted pre-rostral plate (tuberculated plate near right anterior margin, top left in picture). B: AMF.54218, fragment of cranial shield with well-preserved right nasal capsule, ventral view. C: AMF.61373h, isolated fragment of left nasal capsule, ventral view.

near the anterior margin. The anterior lamina is almost vertical and gently convex. The ventral margin of the anterior lamina of the rostral plate is pointed mesially and emarginated laterally immediately anterior to the nasal capsules (Fig. 2A–D). A single pre-rostral plate is preserved in specimen AMF.54614 (Figs. 2C, 3A, 4B), as a small, detached plate lying against the rostral margin of the headshield in front of the right nasal capsule. The plate is symmetrical, wide (7 mm) and very short (1 mm in the midline, 2 mm at either end). One margin is gently concave; the other is deeply indented. The dermal surface is covered with small, closely packed tubercles similar to those present in *Mulgaspis evansorum* cranial and trunk shields. The most likely interpretation is that this little plate, now detached, has not moved far from where it was originally attached – against the anteroventral margin of the rostral plate. It is interpreted here as an internasal, or pre-rostral, plate by comparison with *Kujdanowiaspis* (*Ra*, Stensiö 1969, fig. 8A). Unlike this bone in *Kujdano-wiaspis*, which is depicted as small and narrow, and longer than wide, the pre-rostral of *Mulgaspis* is much wider than long.

This find is significant because a similar crescentic, and tuberculated, pre-rostral plate has been observed by the writer in articulated specimens of *Groenlandaspis* sp. nov. from Mount Howitt, Victoria, Australia, and another possible example has recently been figured by Daeschler *et al.* (2003, fig. 8), in *Turrisaspis elektor* from the Late Devonian of Pennsylvania, although in *Turrisaspis* it is interpreted, probably incorrectly, as the anteroventral margin of the rostral plate. It is suggested here that the presence of a separate, wide and short tuberculated pre-rostral plate is probably characteristic of all Groen-landaspididae, although it is only likely to be detected in unusually well-preserved articulated specimens.

The post-nasal plate in *Mulgaspis* is retained as a separate plate. These elements are lost or fused to the rostral in many other phlyctaeniids. The post-nasal plate of *Mulgaspis* (Figs. 1C, 4C, E, F) is small, subtriangular and steeply inclined, and forms the anterior margin of the orbit. It has a small, unornamented, posteroventrally directed process which was probably the articular surface or overlap area for the suborbital plate, presumably loosely attached to the skull, and not yet identified in the associated material. Note that a reconstruction of the *Mulgaspis* skull roof (?*Mulgaspis evansorum*) drawn by Young & Goujet (2003, fig. 26B) from a cast of a Mount Jack specimen held in Paris, does not show the post-nasal plates.

The central plates of the skull in *Mulgaspis evansorum* (Fig. 4E) are large, seven-sided, and longer than wide (about 45% of the cranial length). They are partly separated anteriorly and posteriorly by the pineal and nuchal plates, respectively, but have a long, straight, common

mesial suture. The pre-orbitals (Fig. 4E) are relatively short and wide with a raised supra-orbital crest. The complete separation of left and right pre-orbitals by the pineal is an unusual feature in a phlyctaeniid, previously recorded only in *Arctolepis* from Spitsbergen (Goujet 1972, fig. 1C).

The post-orbital and marginal plates in *Mulgaspis evansorum* are conventionally developed, but their ventral margins are rounded, tuberculated and deeply emarginated (Figs. 1B, C, 4E), again indicating that any cheek plates present were not firmly attached to the cranial vault. The post-marginal (Fig. 4E) is a small subtriangular plate with a rounded posterolateral angle. A narrow area along its anteroventral margin is indented, but tuberculated, and was presumably covered in life by the larger, loosely attached submarginal plate functioning as an opercular element, essentially as in *Arctolepis* (Goujet 1972, fig. 2), *Dicksonosteus* (Goujet 1975, fig. 6) and *Groenlandaspis*.

The paranuchal plate (Fig. 4E) is short but unusually wide and irregularly shaped. Immediately lateral to the craniothoracic articulation the paranuchal plate has a prominent, steeply inclined process (Fig. 2A, C, H, I), the para-articular process (Fig. 4) with a roughened, "Siebknochen" texture. This process, which fitted inside the anterior margin of the anterior dorsolateral plate under the condyle when the head was depressed, is more strongly developed in *Mulgaspis* than in *Dicksonosteus* (Goujet 1984, figs. 31, 32). The process is variably developed in different species of *Phlyctaenius* (Young 1983, fig. 9). A well-developed para-articular process is also present in *Holonema* (Miles 1971, fig. 29), *Groenlandaspis* (Ritchie 1975, fig. 2A), and all brachythoracids. *Buchanosteus* has a more complex relationship between the para-articular process on the paranuchal plate and a subglenoid process on the anterior dorsolateral plate (White & Toombs 1972), which Jarvik (1980, p. 373, fig. 295) interpreted as a kind of locking device. The process in *Buchanosteus* (Young 1979, fig. 2) is very small compared with *Mulgaspis*.

The articular fossae of *Mulgaspis evansorum* (Fig. 2A, C, H, I) are large, round and deep, situated close to one another and high up under the posterior dorsal margin (Fig. 4B, D, F), to which they are connected (and supported) by the posterior descending lamina of the paranuchal plate. The craniothoracic articulation of *Mulgaspis* is therefore developed basically as in most phlyctaeniids, but differs from that found in *Groen-landaspis* and *Holonema*, in which the articular condyles and fossae are more transversely elongate.

Nasal capsules and neurocranium

These structures were perichondrally ossified, and are preserved as impressions in several specimens. The posterior wall of the nasal capsules (cribrosal bones) of

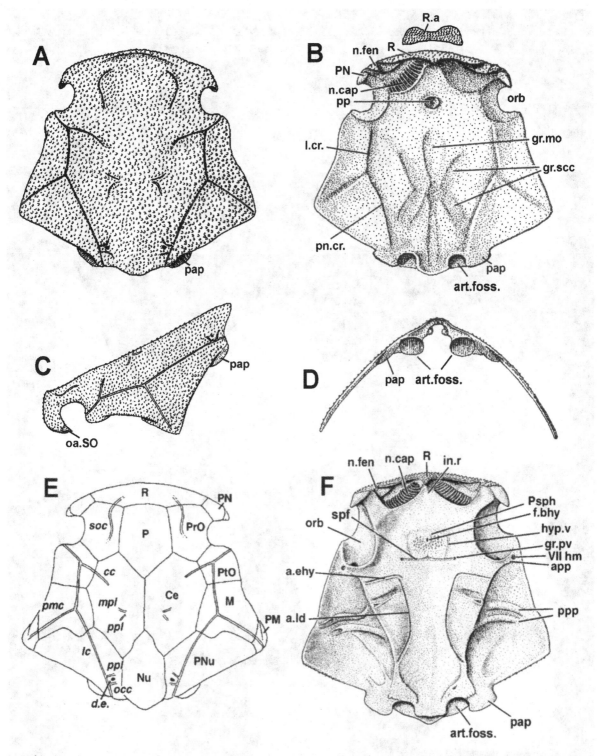

Fig. 4. *Mulgaspis evansorum* gen. et sp. nov., Merrimerriwa Formation, Mulga Downs Group. Restoration of dermal headshield in dorsal (A), ventral (B), lateral (C) and posterior (D) views. E: Pattern of dermal plates and sensory canal grooves, dorsal view. F: Palatal view of skull, with nasal capsules and endocranium restored.

Mulgaspis are preserved, showing that the capsules were anteriorly placed and attached posteriorly to the neurocranium by a continuous post-nasal shelf. The neurocranium consisted of a single platybasic unit, also

perichondrally ossified, running from the occipital region to the ethmoid region. In one specimen of *Mulgaspis evansorum* (AMF.64830A, Fig. 2A, B) the palatal surface was apparently damaged before burial, allowing sediment

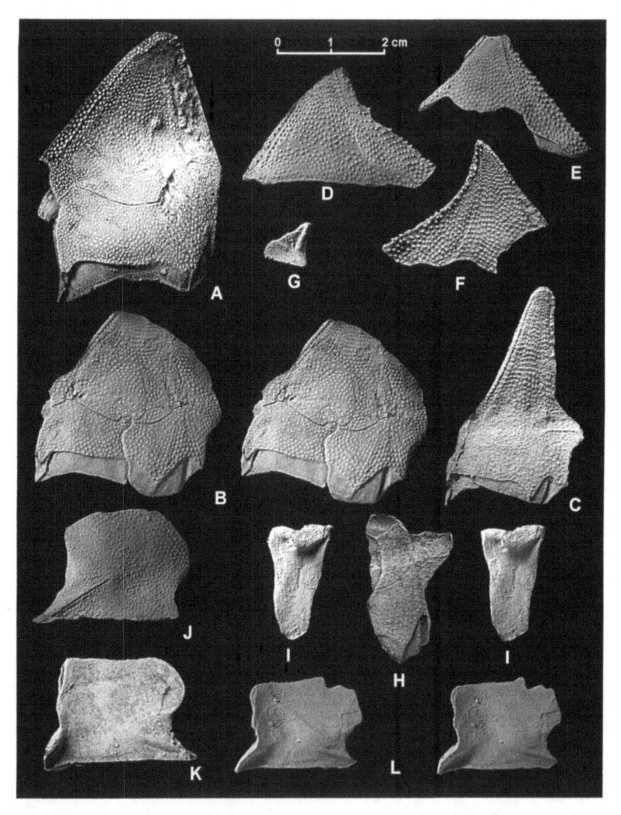

Fig. 5.   *Mulgaspis evansorum* gen. et sp. nov., Merrimerriwa Formation, Mulga Downs Group. A–C: Dorsal plates of trunk shield (median dorsal, anterior dorsolateral and posterior dorsolateral plates) in association, left lateral view. A: AMF.61285B. B: AMF.61365 (stereo-pair). C: AMF.54152. D: Median dorsal plate with parts of anterior and posterior dorsolateral plates, showing common suture (AMF.61369A). E: Median dorsal plate, left lateral view, showing ventral margin (AMF.61619). F: Partial median dorsal plate, right lateral view (AMF.53975). G: Right posterior lateral plate, visceral view (AMF.53975). H: Left posterior dorsolateral plate, external view (AMF.53574). I: Left posterior dorsolateral plate, visceral surface (AMF.54205; stereo-pair). J: Left anterior lateral plate, external surface (AMF.61366). K: Right anterior lateral plate, visceral view (AMF.54162). L: Right anterior lateral plate, visceral view (AMF.65089; stereo-pair).

to infiltrate the endocranial cavity and surround (and preserve) parts of the internal cavities and canals which were lined with perichondral bone.

Stensiö's detailed studies of the complex cranial anatomy of various arthrodires, based on serial sections, provide a firm basis for an interpretation of the *Mulgaspis* braincase. Unlike the European material (mainly *Kujdanowiaspis*) and that from Spitsbergen, most of the specimens from the Mulga Downs Group consist almost entirely of natural moulds. The internal anatomy of the endocranium is only known from a few fortuitously preserved specimens. Thanks to the meticulous, detailed studies and restorations of the endocranium by Stensiö (1969) and Goujet (1984) on phlyctaeniids from the Devonian of the Northern Hemisphere, it is possible to interpret the few specimens from the Southern Hemisphere Mulga Downs Group that display the endocranium, and the similarities are striking.

The dorsal surface of the neurocranium in *Mulgaspis evansorum* can be restored from the impressions left on the undersurface of the headshield (Figs. 1F–H, 4B). The ventral neurocranial surface is preserved in only a few of the many headshields (AMF.54164, 61366, 64830). On the inner surface the orbit is large, wide and deeply concave, reflecting the relatively large size of the eyeball (Fig. 4B). The pineal pit is deep with a raised circular rim. The dorsal neurocranial margins are marked by a prominent ridge, the lateral and paranuchal cristae. A longitudinal groove in the midline for the medulla oblongata is flanked by converging grooves for the anterior and posterior dorsal semicircular canals. The anterior ventral surface (Fig. 2F) is covered by a plexus of small blood vessels radiating from a point just mesial to the orbits; these vessels must have supplied the dorsal surface of the nasal capsules as in *Dicksonosteus* (Goujet 1984, fig. 4, pl.nv.). In *Mulgaspis* the endoskeletal nasal capsule was ossified. The left and right nasal capsules were separated mesially by a smooth ridge and firmly attached anteriorly to the anterior lamina of the rostral plate. In life, this area in *Mulgaspis* was probably covered by the median pre-rostral (or internasal) plate, but much larger and wider than depicted in *Kujdanowiaspis* (Stensiö 1969, p. 88, fig. 8A).

The nasal capsules (cribrosal bones) are ovate, obliquely oriented and open ventrally (Figs. 2A–E, 3A–C). They are about 6 mm across and 4 mm long and the deeply concave dorsal wall is traversed by 16–20 narrow furrows, 2.5 mm long by 0.5 mm wide. These converge anterolaterally on the smooth descending lamina leading to the incurrent nasal fenestra. The fenestra is widest at its inner end but tapers sharply anteriorly to open on the ventral margin below the rostral/post-nasal suture. The concave dorsal wall of the incurrent fenestra is covered with minute tubercles. In one specimen (MF.64830; Fig. 2A, B), the smoothly rounded outer surface of the

posterior wall of the left nasal capsule has been exposed by the loss of part of the ventral surface of the ethmoidal region.

Even in the uncrushed material of *Kujdanowiaspis* the form of the nasal capsule has seldom been found as well preserved as in *Mulgaspis*. Stensiö's (1969, figs. 10A, B, 13) detailed restoration of the shape and structure of the nasal capsule in *Kujdanowiaspis* (and other arthrodires) was based on a few specimens in which the circumcapsular bone was "compressed and distorted" (Stensiö 1945, p. 8) and was "still but slightly known". The perichondral bone lining the roof and walls of the nasal capsule is well preserved in acid-prepared material of *Buchanosteus* and *Parabuchanosteus* (White & Toombs 1972; Young 1979; Young *et al.* 2001), but the new *Mulgaspis* material displays the capsule structure in greater detail than any phlyctaeniid arthrodire to date.

The closely packed ridges on the dorsal and posterior wall clearly must have supported extremely thin lamellae of olfactory epithelium as in modern elasmobranchs like *Scyllium* and *Acanthias* (Grassé 1958, p. 926, fig. 629), and the arthrodire and elasmobranch nasal capsules are remarkably similar in shape (Fig. 3).

The endocranium is platybasic, widest immediately posterior to the orbits across the anterior post-orbital processes (Fig. 4F), and tapers posteriorly to about half this width at the articular fossae. The short anterior post-orbital process encloses the canal for the hyomandibular branch of the facial nerve and a long bifurcating posterior post-orbital process extends from the lateral wall of the neurocranium to the posterolateral angle of the headshield.

The ventral wall of the neurocranium is slightly concave except near the posterolateral margins where it is swollen immediately under the otic regions. Between the orbits is a small ventral foramen surrounded by an ill-defined area of minute tubercles; these represent the buccohypophysial foramen and the parasphenoid, respectively. The parasphenoid (Figs. 2C, 4F) appears to be smaller than that of *Kujdanowiaspis* (Stensiö 1969, p. 88, fig. 8A) and closer to that in *Dicksonosteus* (Goujet 1975, fig. 4) where it is shorter and more crescentic. Behind the parasphenoid is a short, shallow, transverse groove (Fig. 2A, C), the subpituitary fossa (Fig. 4F), disappearing into a pit at either end. A similar groove in *Kujdanowiaspis* (Stensiö 1969, figs. 10B, 57), *Dicksonosteus* (Goujet 1984, fig. 6), and other arthrodires, housed the pituitary vein. Faint curving grooves, running anteriorly from the groove for the pituitary vein on either side of the parasphenoid, may have housed the hypophysial vein as reconstructed in *Dicksonosteus* and *Kujdanowiaspis* (Goujet 1984, figs. 51, 52). This specimen of *Mulgaspis* appears to be the only example to show that the hypophysial veins branched from the pituitary vein canal.

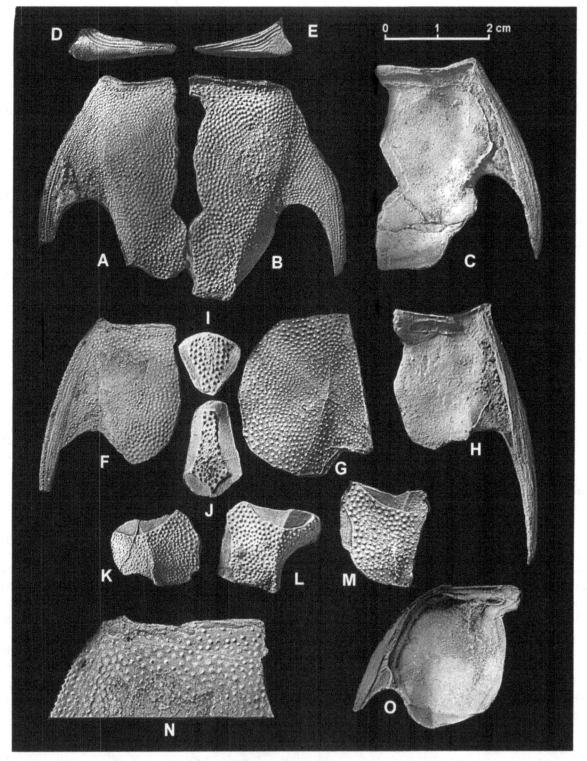

Fig. 6.   Mulgaspis evansorum gen. et sp. nov., Merrimerriwa Formation, Mulga Downs Group. A, B: Half side of ventral trunk shield (anterior ventro-lateral, posterior ventrolateral, anterior ventral, spinal, interolateral plates), external surface (A, right side, AMF.61282; B, left side, AMF.64830). C: Right half of ventral trunk shield (anterior ventrolateral, posterior ventrolateral, spinal, interolateral plates), visceral surface (AMF.65090). D, E: Interolateral plates, anterior view of dorsal lamina (D, right interolateral plate, AMF.64828; E, left interolateral plate, AMF.61282). F, N: Right anterior ventrolateral, anterior ventral, spinal, interolateral plates, external surface (AMF.61294A), with detail (N) showing sutures of the anterior ventral plate between the anterior ventrolateral and interolateral plates. G: Left anterior ventrolateral plate, external surface (AMF.56253). H: Right anterior ventro-lateral, anterior ventral, spinal, interolateral plates, visceral surface (AMF.54695). I: Anterior median ventral plate (AMF.61379D), external view. J: Posterior median ventral plate (AMF.61379G), external view. K–M: Posterior ventrolateral plates, external view (K, AMF.61618E, right plate; L, AMF.61365, left plate; M, AMF.56257, left plate). O: Left anterior ventrolateral, anterior ventral, spinal, interolateral plates, visceral view showing ossified scapulocoracoid.

Two long, curving grooves (Fig. 2A–C) pass forward from the posterolateral margin of the neurocranium, run longitudinally and then turn laterally at right angles to meet the neurocranial margin just posterior to the anterior post-orbital process (Fig. 4F). In *Kujdanowiaspis* a similar groove crosses onto the neurocranium more anteriorly, and passes off ventral (and not posterior) to the anterior post-orbital process (Stensiö 1969, fig. 10B), whereas in *Dicksonosteus* the groove passes off behind the process as it does here. These grooves were interpreted by Stensiö (1969, fig. 57) and Goujet (1984, fig. 51) as having housed the paired cephalic division of the dorsal aorta, and its laterally directed branch, the efferent hyoidean artery (Fig. 4F).

In AMF.64830 (Fig. 2A, B) the left ventral wall of the neurocranium was lost before burial, and some of the internal perichondral ossifications around the otic capsule and associated nerves and blood vessels are partly exposed, but not well preserved. They appear similar to the structures present in *Kujdanowiaspis* (Stensiö 1969, fig. 44).

Trunk shield

No intact examples of the trunk shield of *Mulgaspis evansorum* have been recovered, but all component plates have now been located, in many instances still partly associated. The anterior and posterior dorsolateral plates are most frequently found still firmly attached to the median dorsal plate and only rarely are they found isolated. Similarly, from the ventral shield, the anterior ventrolateral plate is often still bordered by the spinal, interolateral and anterior ventral plates, sometimes with the posterior ventrolateral plate attached. The anterior lateral, posterior lateral and median ventral plates (anterior and posterior) have only been found detached (Figs. 5, 6).

In most of its features, the trunk shield of *Mulgaspis evansorum* is of typically phlyctaeniid-type, relatively long and closed behind the pectoral fenestra, with a narrow median dorsal plate, long spinal plates and a rather flat ventral shield. A well-developed craniothoracic articulation is developed with the condyles set close together. In all of these features the *Mulgaspis* trunk shield resembles phlyctaeniids such as *Dicksonosteus*, *Kujdanowiaspis*, *Arctolepis*, *Phlyctaenius*, etc. (Denison 1978). However, it differs from other phlyctaeniids in two significant aspects, the implications of which are discussed below: (a) the presence of anterior ventral plates and (b) a median dorsal plate which is so narrow and high that the left and right posterior dorsolateral plates meet in the midline inside its base.

The restored trunk shield (Fig. 7) was evidently solidly constructed, triangular in cross-section with a flat or slightly tumid ventral surface, and steeply inclined dorsolateral surfaces terminating in a high, extremely narrow, pointed dorsal ridge. Component bones are illustrated in Fig. 7B, C.

The median dorsal plate had a height/length index of 70–75 and the angle of the pointed dorsal crest was normally around 80°, but some examples could be much more acute (Fig. 5C, F). They range from about 35–40 mm in length but are only 4–5 mm thick at the base. The highest point of the median dorsal plate is centrally placed, the anterodorsal margin is straight or slightly convex, and the posterodorsal margin is straight or concave and usually fringed with a row of prominent tubercles. The anterodorsal ridge is flanked on both sides by a shallow, continuous groove running from the anterior margin to the dorsal crest. The almost vertical flanks of the median dorsal plate are crossed by a steeply inclined low ridge running from the highest point of the posterior dorsolateral plate to the dorsal crest.

The median dorsal plate in *Mulgaspis* is developed essentially like that of *Groenlandaspis*, with a very narrow anterior process and a much deeper, thicker posterior process. The dermal suture between the median dorsal plate and the associated anterior and posterior dorsolateral plates is often indistinct, but is seen clearly in a few well-preserved examples (Fig. 5A, B, D, E). It gradually diverges from the median dorsal plate anterodorsal margin, flattens out over the posterior part of the anterior dorsolateral plate and rises again over the anterior part of the posterior dorsolateral plate. From this point it descends and flattens out under the posterior process of the median dorsal plate.

The anterior and posterior dorsolateral plates have a very complicated and distinctive overlap arrangement with the median dorsal plate and each other, which has proved difficult to determine because these plates are usually found firmly sutured together. In *Groenlandaspis*, the median dorsal plate overlaps the anterior and posterior dorsolateral plates along the full length of their dorsal margin, the dorsolateral plates fitting inside the base of the median dorsal plate. The situation in *Mulgaspis* is quite different.

The posterior dorsal margin of the anterior dorsolateral plate and the anterior dorsal margin of the posterior dorsolateral plate have large well-developed overlap areas for the median dorsal plate, just as in *Groenlandaspis*. The anterior part of the anterior dorsolateral plate dorsal margin lacks such an overlap area, and instead has a small contact face on its visceral surface which meets a small overlap area on the anterior process of the median dorsal plate (Figs. 5E, 7C). This shows that it is the anterior dorsolateral plate which partly overlaps the median dorsal plate and not the reverse, a most unusual occurrence in arthrodires. Similarly the posterior dorsal margin of the posterior dorsolateral plate in *Mulgaspis* lacks an overlap area, but has a contact face on its posterior margin meeting an overlap area on the median dorsal plate (Figs. 5E, 7C). Thus, the posterior dorsolateral plate clearly overlaps

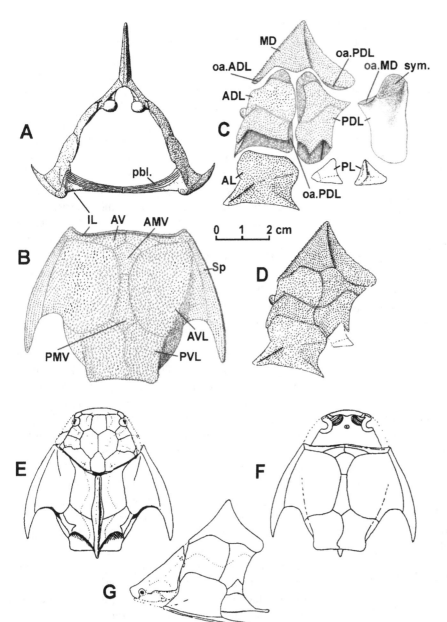

Fig. 7. Mulgaspis evansorum gen. et sp. nov., Merrimerriwa Formation, Mulga Downs Group. A: Trunk shield in anterior view. B: Ventral trunk shield, external view. C: Median dorsal and left lateral trunk plates, illustrating overlap relationships. D: Left lateral trunk plates re-assembled. E–G: Outline restorations of combined head and trunk shields, in dorsal (E), ventral (F), and left lateral (G) views.

the median dorsal plate, again the reverse of the normal arthrodiran condition.

The anterior/posterior dorsolateral plate overlap relationship is also rather unusual (Figs. 5B, H, 7C). In *Mulgaspis* the lateral line sensory groove crosses from the anterior dorsolateral plate to the posterior dorsolateral plate about midway along their common boundary. Above this the anterior/posterior dorsolateral plate suture is concave anteriorly and the anterior dorsolateral plate overlaps the posterior dorsolateral plate, the conventional arrangement. Ventral to the lateral line groove, however, the anterior/posterior dorsolateral plate suture is convex anteriorly, and the posterior dorsolateral plate extends forwards and overlaps the anterior dorsolateral plate, the reverse of the usual condition. The same unusual overlap

relationships of median, anterior dorsolateral and posterior dorsolateral plates are seen in *Mulgaspis altus* sp. nov. described below, and hence are included in the generic diagnosis.

The anterior dorsolateral plate of *Mulgaspis evansorum* is slightly higher than long (height/length index of ornamented area 110–125), S-shaped in cross-section, concave dorsally, and convex ventrally. The articular condyles are large, rounded and set close together. Ventral to the condyle a convex obstantic area with a roughened area on its visceral margin received the para-articular process of the paranuchal plate. The antero-ventral angle of the anterior dorsolateral plate is bluntly pointed and extends ventrally in front of the long, deep overlap area for the anterior lateral plate (Fig. 7C). The lateral line groove crosses onto the anterior dorsolateral plate at the level of the condyle, rises sharply for about 5 mm, with a short dorsal branch (the anterior dorsolateral sensory line), then descends gradually towards the posterior margin. It crosses the anterior/posterior dorsolateral plate border about 45% above the base, much higher than in any described species of *Groenlandaspis* (0–25% above the base). On the posterior dorsolateral plate (Figs. 5A–C, 7C, D) the lateral line groove first curves dorsally, then levels off, the posterior part of the groove running horizontally to the posterior margin, rather than deflected posteroventrally as in most groenlandaspidids. Again this feature is present in *Mulgaspis altus* (described below), and is also proposed as a generic character of *Mulgaspis*.

The anterior and posterior lateral plates have only been found as isolated plates, but from their size, shape and ornament they can be confidently assigned to this species. A few examples (Fig. 5J) display a rounded convex dorsal margin, but in most specimens the dorsal margin is almost straight and parallel to the spinal margin (Fig. 5K, L) and the same profile is clearly seen in the overlap areas on the anterior and posterior dorsolateral plates (Fig. 5A–C). The anterior lateral plate is longer than high (height/length index 68–73), a point of difference from *Mulgaspis altus* sp. nov., where the anterior lateral plate is higher than long. The anterior margin is almost vertical dorsally but projects sharply ventrally, with a strongly inflected infra-obstantic margin where it meets the ascending branchial lamina of the interolateral plate.

Only one example of the posterior lateral plate has been recognised, represented by a mould of the visceral surface (Fig. 5G). Its size, triangular shape and strongly developed ridge on the inner surface confirm this to be a right posterior lateral plate of *Mulgaspis evansorum* which would fit neatly into an overlap area like that seen on the posteroventral margin of the posterior dorsolateral plate (Fig. 5A–C, H).

Ventral shield

Although no complete example of the ventral trunk shield of *Mulgaspis evansorum* has yet been found, all of its component plates have been recovered. Partly associated remains confirm that the ventral shield was relatively short, broad and gently convex (Fig. 7B).The most complete specimens (AMF.61282, 64830; Fig. 6A, B) represent the right and left halves of the shield, respectively. In each, the anterior ventrolateral, posterior ventrolateral, spinal, interolateral and anterior ventral plates are still attached, with only the median ventrals (anterior and posterior) missing. The latter are known from isolated examples (Fig. 6I, J).

The anterior ventrolateral plate is as broad as long, with a gently convex anterior, mesial and posterior margin and an almost straight lateral margin where it meets the spinal plate (Fig. 6F, G). In contrast, the posterior ventrolateral plate is very short and wide, the right posterior ventrolateral plate overlaps the left, and the posterolateral corner of the ventral lamina is quite angular. The index of the length of the anterior/posterior ventrolateral plates is about 200. The spinal plate is long, gently curved and ornamented with fine, parallel tuberculated ridges. The free portion of the spinal plate is most commonly around 25–30% of the total length (Fig. 6A–C, F), but there is considerable variation, with some examples displaying relatively long free spines (Fig. 6H; about 50%) and others with very short spines (Fig. 6O; about 20%). None of the specimens displays any trace of the hook-like spinelets on the inner margin of the free spine that are present in many other phlyctaeniids (e.g. *Dicksonosteus*, *Arctolepis*, and *Groenlandaspis*).

Several examples of the interolateral plate have been recovered in association with the anterior ventrolateral and spinal plates. The ventral lamina of the interolateral plate is narrow, straight and ornamented with small tubercles similar to those over the other ventral plates. The anterodorsally facing post-branchial lamina is subtriangular, very low mesially and rising to about 8 mm high where it meets the spinal and anterior lateral plates. It is ornamented with six to seven finely denticulated ridges, but the area between these and the anterolateral corners lacks ridges and is covered with very fine tubercles. In this feature *Mulgaspis evansorum* differs from *Groenlandaspis*, in which the post-branchial ridges are much more numerous and extend almost to the junction with the spinal plate. A similar feature is seen in *Mulgaspis altus* sp. nov. described below and is considered diagnostic for the genus.

The shape and relative proportion of the anterior and posterior median ventral plates are clearly indicated by the inner margins of the associated right and left ventral shields (Fig. 6A, B). Their overlap areas are also clearly visible on another specimen (AMF.65090) seen in visceral view (Fig. 6C). Examples of both plates have been recovered from Mount Jack (Fig. 6I, J). The anterior median ventral plate is short, subtriangular, and widest anteriorly. The anterior overlap areas are very narrow and gently convex, the posterolateral overlap areas are wider and taper posteriorly. The posterior median ventral plate is longer and narrower, tapering anteriorly, and widest near the posterior margin. The length ratio of the anterior/posterior median ventral plates is about 72.

One of the most unexpected discoveries in *Mulgaspis evansorum* was the presence of small, triangular, paired anterior ventral plates, wedged between the interolateral and anterior ventrolateral plates mesially and meeting in the midline anterior to the anterior median ventral plate (Fig. 6 A, B, F, H). Several specimens (e.g. AMF.61294A; Fig. 6F, N) clearly display both the anterior and posterior sutures of this plate, dispelling any suggestion that it

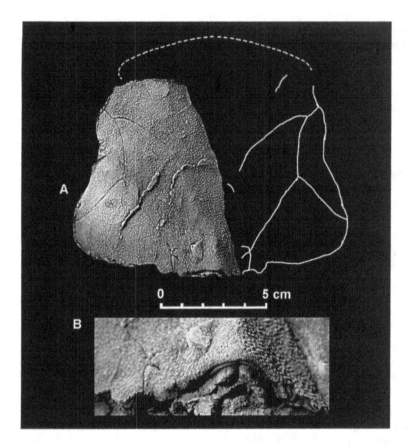

Fig. 8. Mulgaspis altus gen. et sp. nov., Merrimerriwa Formation, Mulga Downs Group. A: Outline restoration from partial headshield (AMF.61323) in external view (Mount Grenfell Station, lower horizon). B: Detail showing the mid-posterior margin of the same specimen in oblique view.

may represent an expansion of the ventral lamina of the interolateral plate.

Paired anterior ventral plates are present in a wide range of actinolepid arthrodires (Denison 1978, figs. 33A–I, 35, *Bryantolepis*; fig. 36, *Sigaspis*), but separate anterior ventral plates have not previously been reported in any phlyctaeniid. There can be no doubt that, with the exception of this character, *Mulgaspis* gen. nov. clearly belongs in the Phlyctaeniidae, not in the Actinolepidae.

Endoskeletal shoulder girdle.

The cartilaginous endoskeletal shoulder girdle (scapulo-coracoid), attached in placoderms to the inner surface of the anterior lateral, spinal, interolateral and anterior ventrolateral plates, is only preserved where it was perichondrally ossified. In the Arthrodira it was long and low, and in actinolepid arthrodires it is best known in *Kujdanowiaspis*? sp. (Stensiö 1969, fig. 188A, B). In the *Mulgaspis* material, several specimens with associated anterior ventrolateral, spinal, interolateral and anterior ventral plates have well-preserved impressions of the scapulocoracoid (Fig. 6C, H, O), confirming that it was constructed essentially like that of *Kujdanowiaspis*.

The scapulocoracoid comprises a transverse anterior portion and a lateral, longitudinal section. Anteriorly a long, narrow coracoid process extends medially inside the

dorsal and ventral laminae of the interolateral plate to meet its antimere in the midline or to terminate just short of it, immediately over the anterior ventral plate. The scapular portion fills the internal cavity of the spinal plate, from the interolateral plate pectoral spine, and extends medially onto the lateral visceral margins of the anterior ventrolateral and anterior lateral plates. Posteriorly the scapulocoracoid extends across the pectoral fenestra where it presumably carried a low transverse crest for the articulation of the pectoral fin endoskeleton (not preserved).

The scapulocoracoid is most often preserved inside the plates of the ventral shield and leaves less trace on the visceral surface of the anterior lateral plate. Its ventral and dorsal surfaces were traversed radially by numerous, bifurcating neurovascular canals. Well-preserved natural moulds of the scapulocoracoid (AMF.65090, 54695; Fig. 6C, H) show the posterior wall of the coracoid process (inside the interolateral plate) with a ridge running inside the ascending dorsal margin of the post-branchial lamina. The inner surface of the scapular section (inside the spinal and anterior ventrolateral plates) is a long shallow groove with many small pits marking the entry point of the neurovascular canals. It also reveals clearly the relationship between the coracoid process and the anterior ventral plate.

***Mulgaspis altus* sp. nov.**
Figs. 8–13

Synonymy. –

1973 "*Wuttagoonaspis fletcheri* (?) spinal plate" – Ritchie, pl. 6, fig. 5

1973 "*Wuttagoonaspis fletcheri*. Anterior ventrolateral, interolateral and part of spinal" – Ritchie, pl. 6, fig. 6

2003 "groenlandaspid gen. et sp. nov." (*in pars*) – Young & Goujet, fig. 30E

Etymology. – Named for the very high and short trunk shield; from the Latin, "*altus*" meaning high.

Diagnosis. – As for genus, with the following additions. Groenlandaspidid with combined head and trunk shield up to about 230 mm long. Headshield up to 105 mm long and about 140 mm wide. Median dorsal ridge with straight anterodorsal and posterodorsal margins. Anterior lateral plate higher than long (height/length ratio approximately 112). Posterior ventrolateral plate much shorter than anterior ventrolateral plate (ventral lamina length ratio about 200). Post-pectoral spine of spinal plate longer than fixed portion.

Remarks. – Several sites in the Dunlops Range are notable for the large size of the arthrodiran plates they contain. Three of these sites have yielded plates assigned here to a relatively large groenlandaspidid, *Mulgaspis altus* sp. nov. Despite a considerable discrepancy in size, and some differences in relative proportions (e.g. of the anterior lateral plates), the most significant diagnostic features of *Mulgaspis evansorum* (dermal ornament, median dorsal plate shape, complex overlaps of median dorsal, anterior dorsolateral and posterior dorsolateral plates, lateral line canal course on the posterior dorsolateral plate, anterior median ventral plate shape indicating the presence of anterior ventral plates, ridged post-branchial lamina on interolateral plate, etc.) all justify placing this second species in the same genus.

Holotype. – AMF.54567, median dorsal plate, Wuttagoona Station (Fig. 9B).

Referred material. – Partial headshields: AMF.61323 (Fig. 8), 64981, 64982; left paranuchal plate (one of two on the same block) F.61315 (Fig. 9A). Trunk plates: median dorsal plates F.54588, F.64986 (juvenile); median dorsal plate with associated left anterior and posterior dorsolateral plates on the same slab, F.64985 (Fig. 10A, C); right anterior dorsolateral plate, F.54571 (Fig. 9D); anterior dorsolateral plate, F.64984; part of left anterior dorsolateral plate with condyle, F.54616 (Fig. 10B); left posterior dorsolateral plate, F.54562 (Fig. 9E); left

anterior lateral plate, F.64983 (Fig. 10E); right anterior lateral plate, F.64987 (Fig. 10D); right anterior lateral plate, F.54617 (Fig. 10F); spinal, F.54680 (Fig. 9F); left anterior ventrolateral, interolateral and spinal plates, visceral view, F.47661 (Figs. 11A, B, 12A–C); right anterior ventrolateral plate, visceral view, F.54573 (Fig. 11D; associated with right anterior dorsolateral plate on the same slab); right posterior ventrolateral plate, F.54564 (Fig. 11E); anterior median ventral plate, F.54616 (Fig. 11C; associated with anterior dorsolateral plate with condyle, Fig. 10B).

Localities. –

(4) Type locality: Wuttagoona Station, 61 km northwest of Cobar, 7 km northwest of Wuttagoona homestead (ruins) and 3 km south of Mount Booroondara (also *Wuttagoonaspis* type locality) F.47661 (anterior ventrolateral, interolateral and spinal plates); F.54588 (median dorsal plate); F.64986 (median dorsal plate).

(5) Tambua Station, 70 km west-northwest of Cobar, NSW, from a low north–south ridge, 5 km north of Tambua homestead and 2–300 m east of Kurrie's Tank. F.54562 (posterior dorsolateral plate) F.54571; F.54616 (anterior median ventral plate); F.54616 (anterior dorsolateral plate); F.64983 (anterior lateral plate); F.64985 (median dorsal plate); F.64987 (anterior lateral plate).

(6) Mount Grenfell Station, 70 km northwest of Cobar; on the crest of a low ridge 8 km north-northwest of Mount Grenfell homestead and 1 km east of the road. F.54564 (posterior ventrolateral plate); F.54567 (median dorsal plate); F.54573 (anterior ventrolateral plate); F.54617 (anterior lateral plate); F.54680 (spinal plate); F.61315 (paranuchal plate); F.64984 (anterior dorsolateral plate).

Horizon and age. – Merrimerriwa Formation, Mulga Downs Group, early Eifelian.

Description. – The material attributed here to *Mulgaspis altus* comes from widely scattered areas on each of three main sites and, with a few exceptions, comes from different individuals.

Headshield

Only a few partial headshields and cranial bones (including several paranuchal plates) have been identified to date as belonging to *Mulgaspis altus* sp. nov., all preserved in external view. The most complete headshield (AMF. 61323, Fig. 8A, B) was recovered on a large slab from a site (not since relocated) described as "about 1 mile south of Mount Grenfell homestead" by the owner, the late Jim Spencer, who donated it to the Australian Museum. This suggests that it comes from the lower horizon in the

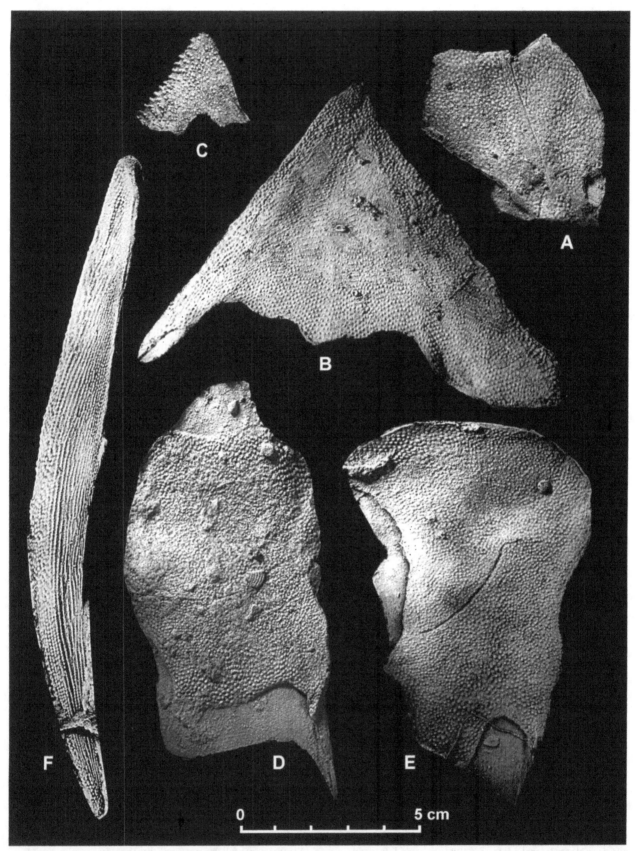

Fig. 9.   *Mulgaspis altus* gen. et sp. nov., Merrimerriwa Formation, Mulga Downs Group. A: Left paranuchal plate (AMF.61315). B: Median dorsal plate, left lateral view (holotype, AMF.54567). C: Small median dorsal plate, right lateral view (AMF.64986). D: Right anterior dorsolateral plate (AMF.54571). E: Left posterior dorsolateral plate (AMF.54562). F: Spinal plate (AMF.54680; originally figured in error as *Wuttagoonaspis* by Ritchie 1973, pl. 6, fig. 5).

Merrimerriwa Formation characterised by a coarser lithology and containing larger placoderm plates – equivalent to the horizons on Wuttagoona Station, Tambua Station (Kurries Tank), and the main site on Mount Grenfell Station that have produced the other *Mulgaspis altus* remains described here.

About a third of the headshield is preserved as an external mould, representing the left posterolateral corner of the skull (Fig. 8A). The ornament consists of fine tubercles. Most of the sensory canal grooves are clearly visible, but there is little or no trace of plate junctions. The canal pattern compares closely with that in *Mulgaspis evansorum* (Fig. 4E), including the very short posterior pitline and occipital commissure, both clearly visible branching off the lateral canal on the paranuchal plate. However, there is no trace of an opening for the ductus endolymphaticus. The original shape, size and proportions of the headshield can be reconstructed suggesting that it was about 10 cm long in the midline and 12–13 cm wide at the posterolateral angles.

Two other partial headshields (AMF.64981, 64982), both natural moulds of the dorsal surface, should be noted here. They were isolated finds from Mount Jack Station (see *Mulgaspis evansorum* above for locality details), and almost certainly belong to *Mulgaspis*, but fall well outside the size range of *Mulgaspis evansorum*, yet are only about half the size of the specimen described above as *Mulgaspis altus*. They are provisionally referred here to *Mulgaspis altus*, pending the recovery of equivalent trunk plates from the Mount Jack sites.

AMF.64982 represents only the anterior half of a headshield. The rostral margin is broad and gently convex. The right orbit is preserved but not the left; sufficient remains to indicate a width of about 48 mm measured across the anterolateral corners of the headshield. Parts of the lateral line groove are visible on both sides with the central and infra-orbital canals visible on the right side. The plate pattern is very indistinct but faint boundaries of the pineal plate clearly show a short, broad, polygonal plate with straight margins, narrowing slightly anteriorly, exactly as in *Mulgaspis evansorum* (Fig. 4E).

AMF.64981 represents most of the right side of a headshield about 64 mm long, with most of the characteristic sensory canal grooves well preserved but again little trace of plate boundaries. The sensory groove system conforms closely to the pattern seen in *Mulgaspis evansorum* (Fig. 4E) and to the large headshield from Mount Grenfell Station also referred to *Mulgaspis altus* (Fig. 8A, B). The lateral line canal (Fig. 4C), central canal, post-marginal canal, posterior pitline and occipital cross-commissure are all visible, together with the opening for the ductus endolymphaticus. Anteriorly the supra-orbital canal curves anteromesially from the pre-orbital plate onto the rostral plate.

Several large, isolated cranial plates from the Dunlops Range sites are also referred here to *Mulgaspis altus*. The more complete of two left paranuchal plates on F.61315 (Fig. 9A) is 53 mm in length and breadth. It lacks the overlap areas but appears to have similar proportions to those of *Groenlandaspis antarctica* and other *Groenlandaspis* spp. In *Mulgaspis evansorum* the paranuchal plate is about one-third the cranial length; if this species has similar proportions, this would indicate a possible length of up to 160 mm for the headshield of *Mulgaspis altus*.

The posterior mesial margin of the paranuchal plate is deeply indented for the lateral process of the nuchal plate which was clearly widest posteriorly as in *Groenlandaspis*. The posterior pitline and occipital cross-commissure meet the lateral line canal groove near the posterior margin (Fig. 9A; cf. Fig. 4E), but there is no clear trace of the opening for the ductus endolymphaticus. A well-developed para-articular process is partly visible on this and the second paranuchal plate on F.61315, but is better exposed on a third example (F.61337), where it is 21 mm long and 9 mm wide, with a concave roughened dorsal surface similar to that of *Groenlandaspis*, *Holonema* and most brachythoracids.

Trunk shield

Many fragments of the median dorsal plate of *Mulgaspis altus* have been recovered, mainly from the type locality on Wuttagoona Station. The more complete examples (Figs. 9B, C, 10A) give a good idea of the distinctive shape and structure of the plate. They range in length from about 37 to 115 mm and in height from about 33 to 97 mm. The high, pointed dorsal crest is centrally situated and the plate is extremely narrow throughout its length; the largest median dorsal plate of *Mulgaspis altus* is only 8 mm thick at the base and tapers dorsally.

The steeply inclined anterodorsal and posterodorsal margins of the median dorsal plate are almost straight and meet at 75°. The sinuous ventral margin, bordering the anterior and posterior dorsolateral plates, is broadly similar to that of the various species of *Groenlandaspis* but with some significant differences. In *Mulgaspis altus* (as in *Mulgaspis evansorum*) these reflect the very unusual overlap relationships of the median dorsal, anterior dorsolateral and posterior dorsolateral plates. The anterior and posterior dorsolateral plates of *Mulgaspis altus* are generally found isolated from the median dorsal plate and were not usually fused, as appears to have been the case in most examples of *Mulgaspis evansorum*. As a result they display the nature and extent of the overlap areas more clearly than the smaller species. The long, narrow, anterior process of the median dorsal plate has a small, recessed, unornamented area along its ventral margin (Fig. 9B), which clearly formed an overlap area for the anterodorsal angle of the anterior dorsolateral plate (Fig. 13A). The much deeper posterior process of the median dorsal plate also displays a low, crescentic, overlap area, in this case for the posterodorsal angle of the posterior dorsolateral plate. In both cases it is the anterior

Fig. 10.   Mulgaspis altus gen. et sp. nov., Merrimerriwa Formation, Mulga Downs Group. A: Median dorsal plate, left lateral view (AMF.64985; same specimen as in C). B: Part of left anterior dorsolateral plate, showing articular condyle (AMF.54616; stereo-pair). C: Partial left anterior and posterior dorsolateral plates showing lateral line groove, and extent of ventral overlap areas (AMF.64985; same specimen as in A). D: Partial left anterior lateral plate (AMF.64987). E: Incomplete left anterior lateral plate (AMF.64983). F: Right anterior lateral plate, visceral view (AMF.54617B).

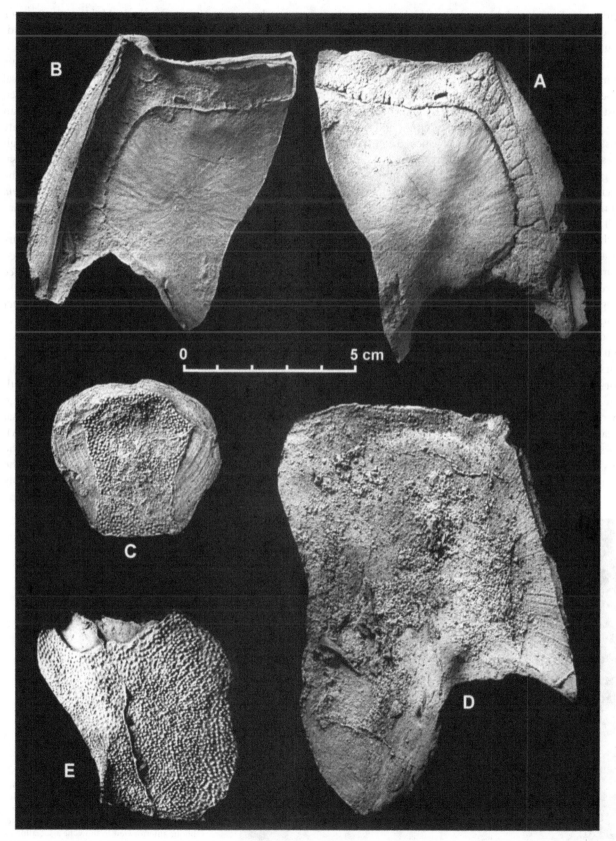

Fig. 11.   Mulgaspis altus gen. et sp. nov., Merrimerriwa Formation, Mulga Downs Group. A, B: Left anterior ventrolateral, anterior ventral, spinal, interolateral plates in association (AMF.47661; originally figured in error as *Wuttagoonaspis* by Ritchie 1973, pl. 6, fig. 6). A: Original specimen showing the natural mould of scapulocoracoid ossification. B: Latex cast in dorsal view. C: Anterior median ventral plate, external surface (AMF.54616). D: Right anterior ventrolateral plate, visceral surface (AMF.54573). E: Right posterior ventrolateral plate, external surface (AMF.54564).

and posterior dorsolateral plates which overlap the median dorsal plate and not the reverse which is the usual situation in arthrodires.

In a few specimens of the median dorsal plate preserved as natural moulds, the sedimentary infilling of the hollow base of the plate takes the form of two triangular natural casts which represent the cavities originally occupied by the high, pointed dorsal overlap areas of the anterior and posterior dorsolateral plates, parts which are not always preserved on the isolated specimens of the latter plates (Fig. 13A). The shape of the anterior and posterior dorsolateral plates internal overlap areas has been restored from these infillings inside the median dorsal plate.

The anterior dorsolateral plate of *Mulgaspis altus* is higher than long (height/length index of ornamented area 145) and tapers dorsally and ventrally. The anterior margin is almost vertical, with a prominent obstantic area developed below the level of the condyle extending into a long, sharply pointed anteroventral angle. The posterior margin is strongly convex dorsal to the lateral line canal groove, and concave ventral to it. The anterior dorsolateral plate thus overlaps the posterior dorsolateral plate dorsally and is overlapped ventrally by the posterior dorsolateral plate, as in *Mulgaspis evansorum*. The lateral line canal crosses the anterior/posterior dorsolateral plate margin just below the midline, but the anterior overlap area on the posterior dorsolateral plate for the anterior dorsolateral plate (Figs. 9E, 13) extends ventrally beyond this level. The lateral line groove curves posterodorsally over the posterior dorsolateral plate before levelling off

towards the posterior margin, as in *Mulgaspis evansorum*, and in contrast to *Groenlandaspis*, where it descends sharply posteriorly.

The dorsal angle of the posterior dorsolateral plate, restored from the ventral margin of the median dorsal plate (Fig. 9B), rises gradually over the anterior part then descends steeply before levelling out under the median dorsal plate posterior process. The median dorsal plate overlap, restricted to the anterior half of the plate, is high and triangular (Fig. 13A). Posteriorly the median dorsal plate is inserted between the posterodorsal angles of the posterior dorsolateral plates, as in *Mulgaspis evansorum*. On its ventral margin the posterior dorsolateral plate has a small, steep overlap area for the anterior lateral plate, and a much larger one for the posterior lateral plate (Figs. 9E, 10C, 13A) but the latter has a smoothly rounded rather than an angular upper margin. There is no trace of the deep vertical groove normally present in *Groenlandaspis* to receive the visceral thickening on the contact face of the posterior lateral plate, although such a groove for the insertion of the posterior lateral plate is well developed in *Mulgaspis evansorum* (Figs. 5B, C, 7C).

The anterior lateral plate of *Mulgaspis altus* is unusually high and short (height/length index 112) compared with that of *Mulgaspis evansorum* which is longer than high (height/length index 68). In *Mulgaspis altus* the dorsal part of the anterior margin is almost vertical (Fig. 13A), with an inturned obstantic area devoid of ornament. The lower part of the anterior margin, separated from the obstantic area by a deep post-branchial notch is sharply

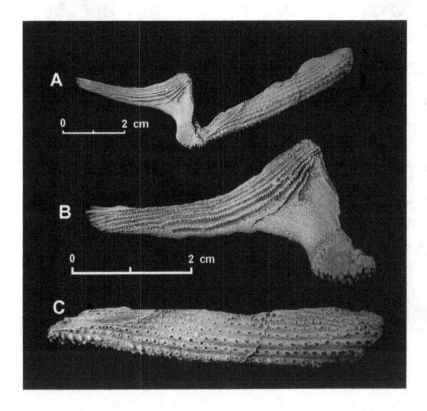

Fig. 12. Mulgaspis altus gen. et sp. nov., Merrimerriwa Formation, Mulga Downs Group. AMF.47661 (see Fig. 11A, B), latex cast made after specimen was split open to expose the interolateral plate and its relationship with the spinal plate. A: Oblique anterior view to show the association of interolateral and spinal plates. B: Detail of the interolateral plate showing a ridged ornament on the internal lamina. C: Detail of ridges, tubercles and pits on the left spinal plate.

inflected and lies under the outer edge of the ascending lamina of the interolateral plate. A low, sharp ridge traverses the long axis of the anterior lateral plate (Fig. 10E) but the other quadrants are not clearly marked. The ornament consists of small tubercles randomly arranged over the middle of the plate but very regularly arranged in oblique intersecting rows near the ventral (spinal) margin. The pectoral embayment is quite deep, and the posterodorsal angle overhangs the rear margin. On its visceral surface (Fig. 10F) there are large, well-developed contact faces for the anterior and posterior dorsolateral plates, and a ventral contact face for the spinal plate which deepens towards the anteroventral angle. There is also an extensive, crescentic roughened area immediately dorsal to the spinal margin and extending well up the pectoral margin; this marks the surface in contact with the dorsal face of the scapulocoracoid.

Material discovered at the same sites from which *Wuttagoonaspis* was described now indicates that two of the trunk plate specimens referred to *Wuttagoonaspis fletcheri* Ritchie (1973, pl. 6, figs. 5, 6) are more likely to belong to *Mulgaspis altus*, which is much rarer in the fauna than *Wuttagoonaspis*. AMF.54680 (Fig. 9F) is an isolated spinal plate, 185 mm long, with a long curved pectoral spine forming about 55% of the total length (compared with *Mulgaspis evansorum* where the pectoral spine is 35–45% of the spinal plate). A second specimen, F.47661 (Fig. 11A, B), consists of the anterolateral part of a left anterior ventrolateral plate with the left interolateral plate and the anterior part of the spinal plate still attached. Part of a genuine *Wuttagoonaspis* spinal, ornamented with very large, rounded, regularly arranged tubercles, can now be recognised lying just posterior to the anterior ventrolateral and spinal plates of *Mulgaspis altus* (Ritchie 1973, pl. 5, fig. 6).

The spinal plate attributed here to *Mulgaspis altus* is exceptionally long, especially in relation to the short but high anterior lateral plate. It is ornamented with many fine, longitudinal ridges crowned with small tubercles. As in *Mulgaspis evansorum*, this spinal plate lacks any trace of hook-like spinelets along the inner surface of the pectoral spine, of the kind commonly found in species of *Groenlandaspis*.

The anterior ventrolateral plate of *Mulgaspis altus*, known only from moulds of the visceral surface (Fig. 11A, B, D), is a large gently bowed plate, up to 114 mm long. It is slightly longer than wide (length/width index about 137), with a straight spinal margin about 75% of the total length. Both specimens lack the mesial margin and the contact faces for the anterior and posterior median

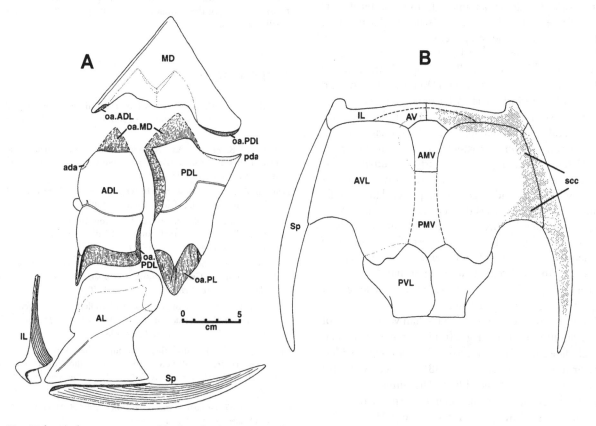

Fig. 13. *Mulgaspis altus* gen. et sp. nov., Merrimerriwa Formation, Mulga Downs Group. A: Restoration of the trunk shield in left lateral view (median dorsal, anterior dorsolateral, posterior dorsolateral, anterior lateral, interolateral, spinal plates), arranged to show the nature of the overlap areas (posterior lateral plate unknown but probably relatively small and subtriangular as in *Mulgaspis evansorum*; Fig. 7C). B: Restoration of the ventral trunk shield with overlap areas indicated by dotted lines on the left, and the extent of the scapulocarocoid shown by the stippled area on the right.

ventral plates. In AMF.54573 (Fig. 11D) the impression of the ventral surface of the scapulocoracoid is clearly visible. Its full extent is best preserved in AMF.47661 (Fig. 11A, B) where a natural mould of the surrounding perichondral ossification reveals it to be similar to that of *Mulgaspis evansorum* (Fig. 6C, H), *Kujdanowiaspis*, and other phlyctaeniids.

Following the recognition that AMF.47661 probably belonged to *Mulgaspis altus*, rather than *Wuttagoonaspis*, the specimen was carefully split and prepared further to reveal a natural mould of the post-branchial lamina of the interolateral plate, in natural association with the anterior portion of the left spinal plate (Fig. 12A, B). The ridged post-branchial lamina is long and low mesially but rises laterally to a pointed crest where it overlaps the inflexed infra-obstantic margin of the anterior lateral plate. The ascending lamina of the interolateral plate is ornamented with six or seven finely serrated ridges (much as in *Mulgaspis evansorum*; Fig. 6D, E). However, the area between the denticulate ridges and the bluntly rounded anterolateral process of the interolateral plate is occupied by an extensive area covered in fine tubercles which extends almost to the interolateral/spinal plate contact. It is thus much less developed than in the interolateral plate of *Groenlandaspis*, where the number of ridges on the post-branchial lamina may reach 12–15, and extend over most of the dorsal lamina right to the anterolateral corner. Otherwise the interolaterals are similar in shape and construction in the two genera.

The posterior ventrolateral plate of *Mulgaspis altus* is known from four examples, all showing only the external surface; the largest, F.54564 (Fig. 11E) is 60 mm long and 60 mm wide anteriorly. The posterior lamina is very short and the posterior margin straight. The right posterior ventrolateral plate overlaps the left mesially, as in *Mulgaspis evansorum*. The lateral lamina of the posterior ventrolateral plate is highest anteriorly, presumably contacting the posterior lateral plate. Posteriorly it descends gradually but it extends right to the posterolateral angle of the plate where it meets the ventral lamina in a short, sharp ventrolaterally directed ridge. Although the posterior ventrolateral plate has not been found in close association with the anterior ventrolateral plate, the largest specimens of both plates suggest that the anterior ventrolateral plate was about 1.9–2.0 times the length of the posterior ventrolateral plate.

There is only one example of a median ventral plate of *Mulgaspis altus*. F.54616 (Fig. 11C) is an anterior median ventral plate found in close association with a partial left anterior dorsolateral plate (Fig. 10B) and may well have come from the same individual. The anterior median ventral plate is roughly pentagonal, the ornamented area is 41 mm long, 35 mm wide anteriorly and 23 mm wide posteriorly. The anterior margin is bluntly rounded, bordered by steep, anterolateral overlap areas. These indicate the probable presence of large well-developed anterior

ventral plates in *Mulgaspis altus*, similar to those found in *Mulgaspis evansorum*. The posterolateral overlap areas for the anterior ventrolateral plates converge posteriorly, but not so sharply as in *Mulgaspis evansorum*. The absence of a posterior overlap area suggests that the anterior median ventral plate overlapped the posterior median ventral plate. The latter must have been considerably longer than the anterior median ventral plate, and experimental re-assembly of the various ventral plates of *Mulgaspis altus* indicates that, with the exception of the anterior ventral plates, it was basically like that of *Groenlandaspis*, but with an unusually wide anterior/posterior median ventral plate contact.

Acknowledgements

The writer acknowledges the hospitality and assistance provided over many years by the Evans family of Tambua Station, and by the late Mr Jim Spencer of Mount Grenfell Station. Assistance in collecting fossil fish material from the widely scattered outcrops of Mulga Downs Group was provided by Kingsley Gregg, Ian Macadie and Robert Jones (all from the Australian Museum). Dr Gavin Young, ANU, Canberra contributed greatly to the completion of the manuscript and provided invaluable information from his discoveries of *Wuttagoonaspis* faunas elsewhere in eastern and northern Australia. I am indebted to Dr Pat Conaghan, Earth and Planetary Sciences, Macquarie University for his contributions to the stratigraphy, correlations and possible dating of the Mulga Downs sequence in the Cobar and Darling Basins.

References

Bembrick, C. 1997: A re-appraisal of the Darling Basin Devonian sequence. *Geological Survey of New South Wales, Quarterly Notes 105*, 1–16.

Daeschler, E.B., Frumes, A.C. & Mullison, C.F. 2003: Groenlandaspid placoderm fishes from the Devonian of North America. *Records of the Australian Museum 55*, 25–60.

Denison, R.H. 1978: *Placodermi. In* Schultze, H.-P. (ed.): *Handbook of Paleoichthyology*, Vol. 5. Gustav Fischer, Stuttgart.

Fletcher, H.O. 1964: New linguloid shells from the Lower Ordovician and Middle Palaeozoic rocks of New South Wales. *Records of the Australian Museum 26(10)*, 283–294.

Glen, R.A. 1979: The Mulga Downs Group and its relationship to the Amphitheatre Group southwest of Cobar. *Quarterly Notes of the Geological Survey of New South Wales 33*, 1–10.

Glen, R.A. 1982a: Nature of late-Early to Middle Devonian tectonism in the Buckambool area, Cobar, New South Wales. *Journal of the Geological Society of Australia 29*, 127–138.

Glen, R.A. 1982b: The Amphitheatre Group, Cobar, New South Wales; preliminary results of new mapping and implications for ore research. *Quarterly Notes of the Geological Survey of New South Wales 49*, 1–12.

Glen, R.A. 1987: *Geology of the Wrightville 1:100,000 Sheet 8034*. New South Wales Geological Survey, Sydney.

Glen, R.A. 1992: Isotopic dating of basin inversion – the Palaeozoic Cobar Basin, Lachlan Orogen, Australia. *Tectonophysics 214*, 249–268.

Goujet, D. 1972: Nouvelles observations sur la joue d'*Arctolepis* (Eastman) et d'autres Dolichothoraci. *Annales de Paléontologie 58*, 3–11.

Goujet, D. 1975: *Dicksonosteus*, un nouvel arthrodire du Devonien du Spitsberg. Remarques sur le squelette visceral des Dolichothoraci. *Colloques Internationaux du Centre National de la Recherche Scientifique 218*, 81–99.

Goujet, D. 1984: *Les poissons placodermes du Spitsberg – Arthrodires Dolichothoraci de la Formation de Wood Bay (Dévonien inférieur)*. Cahiers de Paléontologie (section vertébrés), Editions du Centre Nationale de la Recherche Scientifique, Paris.

Goujet, D. & Young, G.C. 1995: Interrelationships of placoderms revisited. *Geobios, Memoire Special 19*, 89–96

Grassé, P.-P. 1958: Les sens chimiques. *Traité de Zoologie XIII*, 925–930.

Janvier, P. 1996: *Early Vertebrates*. Oxford Science Publication, Clarendon Press

Janvier, P. & Ritchie, A. 1977: Le genre *Groenlandaspis* (Pisces, Placodermi, Arthrodire) dans le Devonien d'Asie. *Comptes Rendues Academie de Science, Paris 284*, 1385–1388.

Jarvik, E. 1980: *Basic Structure and Evolution of Vertebrates*, Vol. 1. Academic Press, London.

Long, J.A. 1984: New phyllolepids from Victoria and the relationships of the group. *Proceedings of the Linnean Society of New South Wales 107*, 263–308.

Long, J.A. 1995: A new groenlandaspidid arthrodire (Pisces: Placodermi) from the Middle Devonian Aztec Siltstone, southern Victoria Land, Antarctica. *Records of the Western Australian Museum 17*, 35–41.

Long, J.A., Anderson, M.E., Gess, R. & Hiller, N. 1997: New placoderm fishes from the Late Devonian of South Africa. *Journal of Vertebrate Paleontology 17(2)*, 253–368.

McCoy, F. 1848: On some new fossil fishes of the Carboniferous period. *Annals and Magazine of Natural History 2*, 1–10.

Miles, R.S. 1971: The Holonematidae (placoderm fishes), a review based on new specimens of *Holonema* from Upper Devonian of Western Australia. *Philosophical Transactions of the Royal Society of London (B) 263*, 101–234.

Miles, R.S. 1973: An actinolepid arthrodire from the Lower Devonian Peel Sound Formation, Prince of Wales Island. *Palaeontographica A 143*, 109–118.

Miles, R.S. & Young, G.C. 1977: Placoderm relationships reconsidered in the light of new ptyctodontids from Gogo, Western Australia. *In* Andrews, S.M., Miles, R.S. & Walker, A.D. (eds): *Problems in Vertebrate Evolution*. Linnean Society Symposium Series 4, 123–198. Academic Press, London.

Mulholland, C. St J. 1940: Geology and underground water resources of the East Darling district. *New South Wales Geological Survey – Mineral Resources 39*, 1–80.

Neef, G., Larsen, D.F. & Ritchie, A. 1996: Late Silurian and Devonian fluvial strata in western Darling Basin, far west New South Wales. *Australian Sedimentologists Group Field Guide Series 10*, 30 pp. Geological Society of Australia, Sydney.

Obruchev, D.V. 1964: *Osnovy paleontologii [Fundamentals in Paleontology. Agnatha, Pisces]*. Nauka, Moscow (in Russian, translation by Israel Program for Scientific Translations, 1967).

Rade, J. 1964: Upper Devonian fish from the Mount Jack area, New South Wales, Australia. *Journal of Paleontology 38*, 929–931.

Ritchie, A. 1969: Ancient fish of Australia. *Australian Natural History 16*, 218–223.

Ritchie, A. 1973: *Wuttagoonaspis* gen. nov., an unusual arthrodire from the Devonian of western New South Wales, Australia. *Palaeontographica A 143*, 58–72.

Ritchie, A. 1974: From Greenland's Icy Mountains – a detective story in stone. *Australian Natural History 18*, 28–35.

Ritchie, A. 1975: *Groenlandaspis* in Antarctica, Australia and Europe. *Nature 254*, 569–573.

Ritchie, A. 2002: A new genus of groenlandaspidid arthrodire (Pisces; Placodermi) from the Early–Middle Devonian Mulga Downs Group of western N.S.W. *In* Brock, G.A. & Talent, J.A. (eds): *First International Palaeontological Congress, 2002, Macquarie University, Sydney, Australia*. Geological Society of Australia, Abstracts 68, 137.

Schultze, H.-P. 1984: The headshield of *Tiaraspis subtilis* (Gross) [Pisces, Arthrodira]. *Proceedings of the Linnean Society of New South Wales 107*, 355–365.

Spence, J. 1958: The geology of the Murray River Basin. Frome-Broken Hill. Unpublished report no. 7500-6-27.

Stensiö, E.A. 1945: On the heads of certain arthrodires. 2. On the cranium and cervical joint of the Dolichothoraci (Acanthaspida). *Kunglica Svenska Vetenskapsakademiens Handlingar 22(1)*, 1–70.

Stensiö, E.A. 1969: Elasmobranchiomorphi Placodermata Arthrodires. *Traité de Paléontologie 4(2)*, 1–692.

Turner, S., Jones, P.J. & Draper, J.J. 1981: Early Devonian thelodonts (Agnatha) from the Toko Syncline, western Queensland, and a review of other Australian discoveries. *BMR Journal of Australian Geology and Geophysics 6*, 51–69.

White, E.I. & Toombs, H.A. 1972: The buchanosteid arthrodires of Australia. *Bulletin of the British Museum of Natural History (Geology) 22*, 370–410.

Woodward, A.S. 1891: *Catalogue of the Fossil Fishes in the British Museum (Natural History), Cromwell Rd, London. Pt. II. Elasmobranchii*. Trustees, British Museum of Natural History.

Young, G.C. 1979: New information on the structure and relationships of *Buchanosteus* (Placodermi, Euarthrodira) from the Early Devonian of New South Wales. *Zoological Journal of the Linnean Society 66*, 309–352.

Young, G.C. 1980: A new Early Devonian placoderm from New South Wales, Australia, with a discussion of placoderm phylogeny. *Palaeontographica A 167*, 10–76.

Young, G.C. 1981: Biogeography of Devonian vertebrates. *Alcheringa 5*, 225–243.

Young, G.C. & Goujet, D. 2003: Devonian fish remains from the Dulcie Sandstone and Cravens Peak Beds, Georgina Basin, central Australia. *Records of the Western Australian Museum, Supplement 65*, 1–85.

Young, G.C., Lelièvre, H. & Goujet, D. 2001: Primitive jaw structure in an articulated brachythoracid arthrodire (placoderm fish; Early Devonian) from southeastern Australia. *Journal of Vertebrate Paleontology 21*, 670–678.

Young, G.C., Long, J.A. & Turner, S. 1993: Faunal lists of Eastern Gondwana Devonian macrovertebrate assemblages. *In* Long, J.A. (ed.): *Palaeozoic Vertebrate Biostratigraphy and Biogeography*, 246–251. Belhaven Press, London.

Young, G.C. & Turner, S. 2000: Devonian microvertebrates and marine–nonmarine correlation in East Gondwana: overview. *Courier Forschungsinstitut Senckenberg 223*, 453–470.

Young, V.T. 1983: Taxonomy of the arthrodire *Phlyctaenius* from the Lower or Middle Devonian of Campbelltown, New Brunswick, Canada. *Bulletin of the British Museum of Natural History (Geology) 37(1)*, 1–35.

A new omalodontid-like shark from the Late Devonian (Famennian) of western Siberia, Russia

ALEXANDER IVANOV & OLGA RODINA

Ivanov, A. & Rodina, O. **2004 06 01**. A new omalodontid-like shark from the Late Devonian (Famennian) of western Siberia, Russia. *Fossils and Strata*, No. 50, pp. 82–91. Russia. ISSN 0300-9491.

Teeth belonging to a new chondrichthyan, *Siberiodus mirabilis* gen. et sp. nov., are described from the Famennian Pescherka and Podonino Formations of the Kuznetsk Basin, and the Cheybekkiol' Formation of Gorniy Altay, western Siberia. The Pescherka Formation is early Famennian based on conodonts (*crepida–marginifera* conodont zones), and the Podonino Formation is late Famennian (*trachytera–expansa* zones). The Cheybekkiol' Formation is also early Famennian (Middle *triangularis*–Early *rhomboidea* zones). The teeth of the several omalodontid-like taxa are characterised by a base with a labial extension, and a reduced lingual part, and most show asymmetry of the crown. The new taxon differs from other such forms in the external and histological structures of the crown, and the type of vascularisation system. It is suggested that chondrichthyan teeth with a labial base extension could have arisen separately in several shark groups.

Key words: Chondrichthyes; shark teeth; new genus *Siberiodus*; Late Devonian; Siberia.

A. Ivanov [aoi@AI1205.spb.edu], Department of Palaeontology, St. Petersburg University, 16 Liniya 29, St. Petersburg 199178, Russia

Olga Rodina [rod@uiggm.nsc.ru], Laboratory of Palaeozoic Palaeontology and Stratigraphy, Institute of Petroleum Geology, Siberian Branch of Russian Academy of Sciences, 3 Prospect Akademika Koptyuga, Novosibirsk 630090, Russia

Introduction

Famennian vertebrates are still poorly known in western Siberia. There have been rare mentions of their occurrence in the Kuznetsk Basin and Gorniy Altay (Ivanov *et al.* 1992; Ivanov & Rodina 2001). However, recently abundant vertebrate assemblages were found in the Famennian of both regions of the Altay-Sayani Folded Belt. Faunal lists for these areas are given in Tables 1–3. This paper describes a new taxon, *Siberiodus mirabilis* gen. et sp. nov., based on unusual chondrichthyan teeth in samples from the Pescherka and Podonino Formations of the Kuznetsk Basin, and another from the Cheybekkiol' Formation of Gorniy Altay, western Siberia.

Locality information

The new taxon described below occurs in some samples from the Kuznetsk Basin: E-9014-19/3 and E-9014-21/3; as well as Gorniy Altay: G 8027a and G 8027b. The section

E-9014 is located on the left bank of the Yaya River (Fig. 1A, B). Sample E-9014-19/3 was collected from the upper part of the Pescherka Formation, represented by greyish-green clay-rich limestones with abundant brachiopods corresponding to the *crepida–marginifera* conodont zones (Yolkin *et al.* 1997). Sample E-9014-21/3 comes from the upper part of the Podonino Formation, which is characterised by yellowish-green clayey brachiopod limestones comprising the *trachytera–expansa* conodont zones (Yolkin *et al.* 1997; E. A. Yolkin, pers. comm.). These samples were probably collected from the *marginifera* and *expansa* zones, respectively (Fig. 2).

Section G 8027 in southeastern Gorniy Altay is exposed in the Chuya River Basin, on the western bank of Cheybekkiol' Lake, 6 km to the north of Aktash village (Fig. 1C). Samples G 8027a and G 8027b are reported from the grey clay-rich limestones with abundant invertebrates (brachiopods, bryozoans, crinoids, etc.) from the upper part of the Cheybekkiol' Formation (Fig. 2). The formation represented by the interbedding of siltstones,

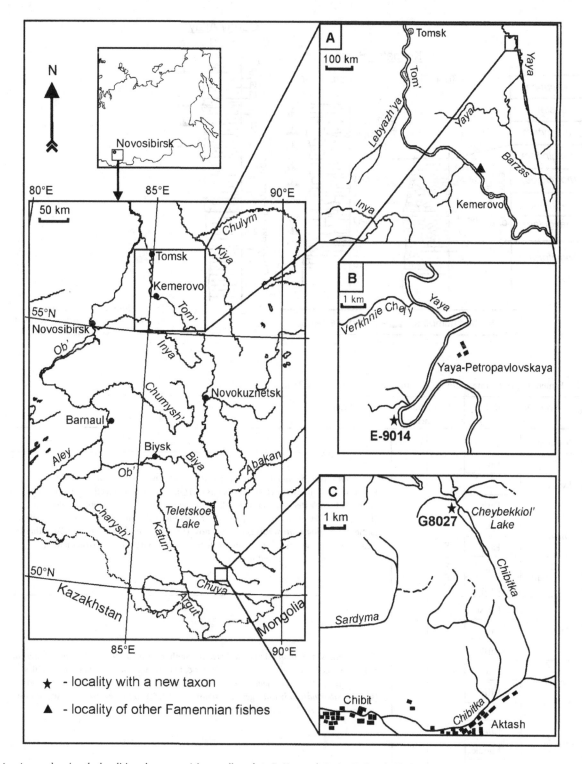

Fig. 1. A map showing the localities where material was collected. A, B: Kuznetsk Basin. C: Gorniy Altay.

sandstones and fossiliferous limestones is correlated with the Middle *triangularis*–Early *rhomboidea* conodont interval (Gutak *et al.* 2001).

Type specimens of the new taxon are housed in the Central Siberian Geological Museum of the United Institute of Geology, Geophysics and Mineralogy in Novosibirsk (prefix CSGM, collection number 838). Further illustrated specimens of *Omalodus* are in the Palaeontological Museum of St. Petersburg University (PM SPU 7).

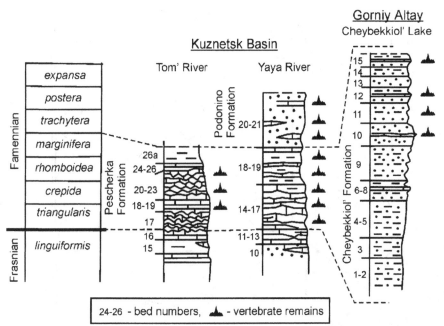

Fig. 2. Stratigraphic sections of Famennian formations, showing the distribution of vertebrate samples.

Table I.   Famennian vertebrate assemblages of the Yaya River, Kuznetsk Basin.

Stage and conodont zones			Formation and member		Vertebrates
	Praesulcata				
	Expansa	Late			*Stethacanthus* cf. *S. thomasi* (Turner) (teeth), Stethacanthidae (teeth), Symmoriida ("Stemmatias"-like denticles), *Phoebodus rayi* Ginter & Turner (teeth), *Phoebodus turnerae* Ginter & Ivanov (teeth), *Phoebodus* cf. *P. typicus* Ginter & Ivanov (tooth), *Jalodus australiensis* (Long) (teeth), Ctenacanthidae (scales), Protacrodontidae (teeth and scales), "*Ohiolepis*" (scales), **Siberiodus mirabilis gen. et sp. nov.** (teeth), "*Acanthodes*" (scales), Acanthodii (scales), Onychodontidae (teeth and scale fragments), Sarcopterygii (fragments), Palaeonisciformes (teeth and scales).
		Middle			
		Early			
	Postera	Late	*Podonino*		
		Early			
	Trachytera	Late			
		Early			
	Marginifera	Latest		Upper	*Stethacanthus* cf. *S. thomasi* (Turner) (teeth), Stethacanthidae (teeth), *Phoebodus rayi* Ginter & Turner (teeth), *Phoebodus turnerae* Ginter & Ivanov (teeth), Ctenacanthidae (scales), *Protacrodus* cf. *P. vetustus* Jaekel (teeth), Protacrodontidae (teeth and scales), ?Hybodontidae (scales), **Siberiodus mirabilis gen. et sp. nov.** (teeth), "*Acanthodes*" (scales), Acanthodii (tooth-like cone), Palaeonisciformes (teeth and scales).
		Late	*Pescherka*		
		Early			
	Rhomboidea	Late			
		Early			
	Crepida	Latest			
		Late			
		Middle			
		Early			
Famennian	*Triangularis*	Late		Lower	Arthrodira (plate fragments), *Stethacanthus* (teeth), Stethacanthidae (teeth), Symmoriida (tooth), *Phoebodus* (tooth fragments), Ctenacanthidae (scales), *Protacrodus* (teeth), Protacrodontidae (teeth), "*Acanthodes*" (scales), Onychodontidae (teeth and scales), Osteolepididae (scales), Sarcopterygii (cf. Rhizodontida, scale), *Moythomasia* (scales), Palaeonisciformes (teeth and scales).
		Middle			
		Early			

Table II.   Famennian vertebrate assemblages of the Tom' River, Kuznetsk Basin.

Stage and conodont zones			Formation and beds		Vertebrates
Famennian	Trachytera		Pescherka		
	Marginifera	Latest		Mitikha	*Stethacanthus* cf. *S. thomasi* (Turner) (teeth), *Phoebodus rayi* Ginter & Turner (teeth), *Protacrodus aequalis* Ivanov (teeth), *Protacrodus* (teeth)
		Late			
		Early			
	Rhomboidea	Late			
		Early			
	Crepida	Latest			
		Late			
		Middle			
		Early			
	Triangularis	Late		Kosoy Utes	*Rhynchodus* (tooth plates), Ptyctodontidae (tooth plate fragments), Holonematidae (plates), Arthrodira (plate fragments), *Stethacanthus* (teeth), Stethacanthidae (teeth), Symmoriida (tooth), *Phoebodus typicus* Ginter & Ivanov (teeth), Ctenacanthidae (scales), *Protacrodus aequalis* Ivanov (teeth), *P.* cf. *P. vetustus* Jaekel (tooth), *Protacrodus* (teeth), Protacrodontidae (teeth and scales), "*Devononchus*" (scales), "*Acanthodes*" (scales), *Holoptychius* (scales), *Kentuckia* (scales), *Moythomasia* (scales), Palaeonisciformes (teeth and scales).
		Middle			
		Early			

Table III.   Famennian vertebrate assemblage from Cheybekkiol' Lake, Gorniy Altay.

Stage and conodont zones			Formation	Vertebrates
Famennian	rhomboidea	Late	Cheybekkiol'	
		Early		Ptyctodontida (fragments), *Stethacanthus* cf. *S. thomasi* (Turner) (teeth), Ctenacanthidae (scales), Protacrodontidae (scales), ?Hybodontidae (scales), ***Siberiodus mirabilis* gen. et sp. nov.** (teeth), "*Devononchus*" (scales), "*Acanthodes*" (scales), Onychodontidae (teeth and scale fragments), Dipnoi (scale fragments), *Moythomasia* (scales), Palaeonisciformes (teeth and scales).
	crepida	Latest		
		Late		
		Middle		
		Early		
	triangularis	Late		
		Middle		
		Early		

Systematic palaeontology

Class Chondrichthyes, Huxley, 1880

Subclass Elasmobranchii Bonaparte, 1838

Order *incertae sedis*

Genus *Siberiodus* nov.

Type species. – *Siberiodus mirabilis* sp. nov.

Etymology. – From Siberia.

Diagnosis. – Teeth with the asymmetrical diplodont crown including large lateral cusps and labiobasal-directed base.

Crown with three up to five separated cusps completely ornamented by straight cristae. Cusps curved linguad, rounded in the cross-section; three larger cusps are unequal in size. Base with strongly reduced lingual and flat basal parts forming an obtuse angle with crown. Vascular canals open on lingual rim and labial side of base as horizontal rows and form a complex network. Cusps consist of orthodentine and enameloid.

***Siberiodus mirabilis* sp. nov.**
Figs. 3, 4, 5A–C

Etymology. – Latin *mirabilis* = extraordinary.

Holotype. – Specimen CSGM 838/1, an isolated tooth (Fig. 3A–E).

Other material. – Five isolated almost complete teeth and nine another fragments, Yaya River, Kuznetsk Basin, sample E-9014-19/3, upper part of Pescherka Formation, possibly the *marginifera* conodont zone; sample E-9014-21/3, upper part of Podonino Formation, possibly the *expansa* conodont zone; Cheybekkiol' Lake, western bank, 6 km to north from Aktash village, Chuya River Basin, Gorniy Altay, samples G 8027a and G 8027b, upper part of Cheybekkiol' Formation, Middle *triangularis–* Early *rhomboidea* conodont zones.

Type locality. – Yaya River, left bank, Kuznetsk Basin; sample E-9014-21/3.

Occurrence and age. – Famennian of Kuznetsk Basin and Gorniy Altay, Russia. The age of the type horizon,

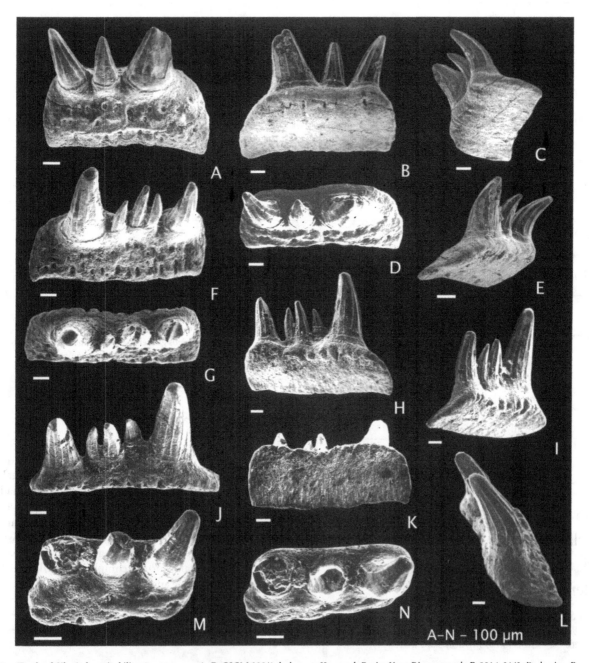

Fig. 3.   Teeth of *Siberiodus mirabilis* gen. et sp. nov. A–E: CSGM 838/1, holotype, Kuznetsk Basin, Yaya River, sample E-9014-21/3, Podonino Formation, probably *expansa* conodont zones. F–L: CSGM 838/2, Kuznetsk Basin, Yaya River, sample E-9014-21/3, Podonino Formation, probably *expansa* conodont zone. M, N: CSGM 838/3, Gorniy Altay, Cheybekkiol' Lake, sample G 8027a, Cheybekkiol' Formation, Middle *triangularis–*Early *rhomboidea* conodont zones. A, F: Labial views. M: Oblique labial view. B, H: Lingual views. G, D, N: Occlusal views. K: Basal view. L: Lateral view. C, E, I: Oblique lateral views. Scale bar = 100 μm.

in the upper part of the Podonino Formation, possibly correlates with the *expansa* conodont zone (late Famennian).

Diagnosis. – As for genus.

Remarks. – A similar tooth from the Frasnian of Iran was described by Janvier (1977) as ?*Cladodus* sp. Unfortunately, only the labial side of this tooth was illustrated, and so this specimen cannot be put into synonymy here. The features by which the new taxon differ from other omalodontid-like teeth are dealt with in the Discussion.

Description. – The teeth range in size from 0.6 to 1.4 mm along the base. They have an asymmetrical diplodont crown. The crown consists of three to five cusps of which the three large cusps are of unequal size, and the two smaller are intermediate cusplets. One of the lateral cusps is one-third of its height higher than the other, and the second lateral cusp is somewhat larger than the central cusp. The cusps of tricuspid teeth are not significantly different in the size of the cross-section near the base. The central cusp and the intermediate cusplets in the multicuspid crown have similar diameters, and are much narrower than the lateral cusps. The cross-section of the cusps is circular or slightly labiolingually compressed. The cusps are separated from each other and do not form a single crown. The lateral cusps either diverge from the axis of the central one or they are almost parallel. All cusps are curved, inclined lingually. They are covered by distinct straight cristae on the whole cusp surface, the cristae reach the top of the cusp, and their number varies from three to five on both lingual and labial sides. The centres of the main cusps are placed on one line but the intermediate cusplets could be placed to labiad.

The boundary of the cusps with the base is very distinct, with a thin groove. The angle between the crown and the base is nearly 120°. The base shows a labiobasal direction and a reduced lingual part. It is elongated mesiodistally, from subrectangular to trapezoid in shape and wider than the crown. The labial side of the base is slightly convex, the basal side is flat.

One horizontal row of up to 12 openings of vascular canals is located on the lingual rim; the labial surface of the base bears two rows of canal openings and some openings between them. The basal side lacks openings. The vascularisation system shows a very complex organisation of canals (Figs. 4E, 5B). It comprises numerous narrow canals, of almost equal diameter, which form a divaricated network filling the whole base. The pulp canal is wide in the lower part, and sharply narrows to the top of the cusp. The canal network of teeth with five cusps is more complicated and ramified. The cusps are composed of an orthodentine with short, slightly

branched or straight dentine tubules (Fig. 4B, D, F). The orthodentine is covered by some thin layers of enameloid.

The arrangement of teeth in the tooth row (base to base) can be reconstructed such that the part of the basal side under the crown of the outside tooth overlapped the distal part of the labial surface at the next younger tooth (Fig. 5C). In this arrangement the row of openings on the lingual rim is positioned opposite the outer row on the labial side.

The teeth vary in the number of cusps (from three to five), in the degree of cusp curvature (from slightly to greatly curved), in their divergence from the central axis (from divergent to almost parallel), in the shape of the base (from subrectangular to trapezoid), and in the ramification of vascular canals (from moderate in tricuspid teeth to intense in teeth with five cusps).

Discussion

Teeth with a lingual extension of the base occur in many groups of Palaeozoic sharks, but several Devonian taxa have the teeth with extended labial and undeveloped lingual parts of the base. These forms were all described from isolated teeth. Turner (1997) erected the order Omalodontida and included some such chondrichthyans based on the shared mentioned characters. *Omalodus*, *Doliodus*, *Portalodus*, and *Siberiodus* possess a similar base. Two taxa, *Aztecodus* and *Anareodus*, have some resemblance to that group. However, our observations demonstrate their essential differences (summarised in Table 4).

Omalodus was erected by Ginter & Ivanov (1992), based on *Phoebodus? bryanti* Wells. Turner (1997) referred *Dittodus grabaui* Hussakof & Bryant to *Omalodus*, and suggested that these two species might be synonymised. There is large variation in the crown of *Omalodus* teeth, but less in the base, and the status of species can only be determined after a complete revision of all material. *Omalodus* has been recorded in the Late Givetian of Kuznetsk Basin, Russia, Poland, Mauritania, Morocco, central and eastern USA (Turner 1997; Ivanov & Derycke 1999; Ginter & Ivanov 2000; Hampe & Aboussalam 2002).

The teeth of *Omalodus* have a mainly symmetrical, phoebodont-like crown, with three main cusps which are almost equal in size, or with the central one a little smaller. The number of intermediate cusplets varies from two to four. The cusps are smooth, slightly sigmoidal as in the most phoebodont crowns, inclined lingually, and circular in cross-section; a lateral crista often forms a web between the cusps. The base has a labiobasal direction, an undeveloped lingual, an extended and convex labial, and concave basal parts. The base forms a large angle

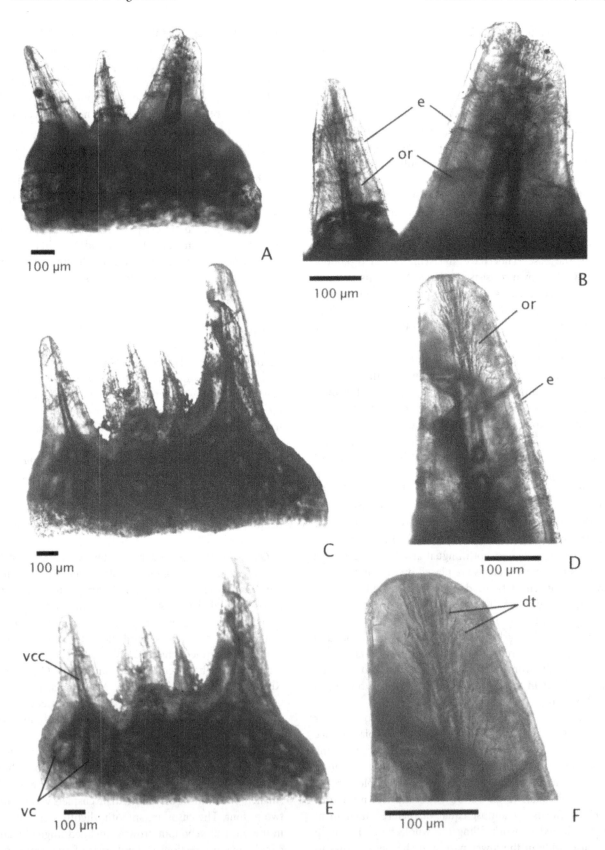

Fig. 4. *Siberiodus mirabilis* gen. et sp. nov., histological structure of teeth immersed in aniseed oil. A, B: CSGM 838/1 (holotype). C–F: CSGM 838/2. dt, dentine tubules; e, enameloid; or, orthodentine; vc, vascular canal; vcc, vascular canal of cusp.

Table IV.   Comparative features of omalodontid-like teeth.

Taxon	Tissue of crown	Crown shape	Number of cusps	Cusp ornamentation	Labial side of base	Foramina of canals on	
						labial side	lingual side
Aztecodus	pleuromin	asymmetrical "diplodont"	2	smooth	reduced	2 openings	2 openings
Doliodus	orthodentine	asymmetrical diplodont	3–6	smooth	extended	?	?
Omalodus	orthodentine + osteodentine	symmetrical phoebodont	3–5	smooth	extended	row in groove	1–3 openings
Portalodus	enameloid + orthodentine + osteodentine	asymmetrical "diplodont"	2	slender striae	extended	row	1–2 openings
Siberiodus	enameloid + orthodentine	asymmetrical diplodont	3–5	distinct striae	extended	2 rows	row

(up to 160°) with the crown. A row of canal openings is located in a longitudinal groove on the labial side at the crown/base boundary (Fig. 5D). There are one to three openings in the centre of the basal side, near the mentioned boundary (Fig. 5E). Vascular canals run across the base from the labial to lingual sides and arise into the cusps (Fig. 5G). The crown lacks enameloid and comprises orthodentine in the most part of the cusp; osteodentine is present in the basal part.

The genus *Doliodus* was established by Traquair (1893) for *Diplodus problematicus* Woodward from the Pragian–Emsian Atholville Beds, Cambellton Formation of New Brunswick, Canada. For a long time the taxon was interpreted as acanthodian (e.g. Denison 1979), but it has been recently re-assigned to the chondrichthyans (Turner & Campbell 1993; Turner 1995, 1998; Miller *et al.* 2003). *Doliodus* teeth, according to these authors, and Woodward's illustration (1892), have the asymmetrical diplodont crown with two large, widely divergent lateral cusps, and two to four intermediate cusplets. The cusps are composed of orthodentine (Miller *et al.* 2003). The base is extended labially, is concave on the basal side, and has a row of large foramina. The teeth form tooth families with connecting bases that are not observed in any other omalodontid-like taxa.

Aztecodus, *Anareodus* and *Portalodus* were reported from the Middle Devonian Aztec Siltstone of Victoria Land, Antarctica (Long & Young 1995; Hampe & Long 1999). All of these forms have teeth with an asymmetrical bicuspid crown, and an undeveloped lingual part of the base. However, they display several differences in the shape of the base and the histological structure of the crown. *Aztecodus* teeth (Fig. 5I) possess a mesial crenulation in the crown which consists of pleuromin without enameloid (Hampe & Long 1999). The base is almost symmetrical, with both lingual and labial parts slightly developed, but with the labial part a little longer, and with two pairs of vascular canals on the lingual and labial faces.

The crown of *Anareodus* teeth is strongly asymmetrical, bearing one very high lateral cusp, a short crenulated area and a tiny cusplet alongside the larger cusp. The base is low and short with a slightly extended labial part. The teeth of *Portalodus* (Fig. 5F) are characterised by the coarse cusp ornamentation on the lingual side, the extension of the labial side of the roundish base, the presence of a row of canal openings on the labial side, and one or two openings on the lingual margin. The crown comprises osteodentine and orthodentine covered by enameloid according to Hampe & Long (1999).

As noted by Turner 1998, the teeth of *Portalodus* are most similar to *Omalodus* teeth in the base shape and direction. We can also add their similarities in the position of canal openings and the involvement of osteodentine in the crown. However, they differ in the external and internal crown structure. *Omalodus* possessed a well-developed phoebodont-like crown, with sigmoidal cusps having a lateral crista, features not found in the other taxa. The *Siberiodus* teeth described here probably most resemble those of *Doliodus* in the cusp arrangement in the crown. The difference of *Siberiodus* teeth from other omalodontid-like teeth are the following: the separated cusps with well-developed ornamentation of strong ridges, the rows of openings on both sides of the base, the complex vascularisation system, and the absence of osteodentine in the crown. *Anareodus* teeth have a less developed labial extension of the base than in the other observed four taxa. *Aztecodus* differs considerably from the other taxa by features such as the crenulation in the crown, the pleuromin composition of the crown, the shape of the base, and the presence of only two canal openings on the labial side.

In summary, the teeth of the taxa just discussed, including *Siberiodus* gen. nov., show a varying degree of labial extension of the base, an undeveloped lingual part of the base, a lack of elements for the tooth-to-tooth articulation in a row, and (except for *Omalodus*) asymmetry of the crown. However, the main differences

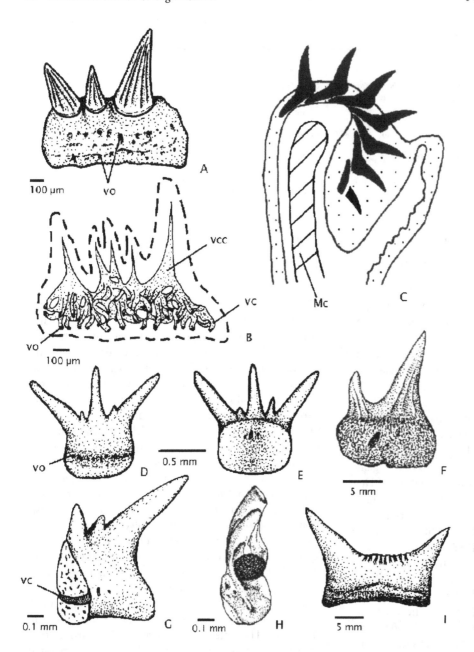

Fig. 5. Various taxa with omalodontid-like teeth. A–C: *Siberiodus mirabilis* gen. et sp. nov. A: Tooth in labial view, CSGM 838/1. B: Reconstruction of vascularisation system, after specimen CSGM 838/2. C: Reconstruction of teeth articulated in a tooth row (cross-section through the jaw after Hampe 1989). D, E, G, H: Teeth of *Omalodus bryanti* (Wells). D, E: Reconstructed tooth in labial and lingual views, based on various specimens. G: Tooth in oblique lateral view with transversal cross-section, PM SPU 7-6. H: Tooth in lateral view, PM SPU 7-4 (after Ginter & Ivanov 1996). F: Tooth of *Portalodus bradshawae* Long & Young, 1995 in lingual view. I: Tooth of *Aztecodus harmsenae* Long & Young, 1995 in labial view (F, I after Hampe & Long 1999). Mc, Meckel's cartilage; vc, vascular canal; vcc, vascular canal of cusp; vo, openings of vascular canals.

lie in the external and histological structures of the crown, and the type of vascularisation system. This suggests that such chondrichthyan teeth, characterised by a labial base extension, could have arisen separately in several shark groups as well the teeth wish a lingually directed base have been appeared in different groups of chondrichthyans.

Acknowledgements

We are most grateful to Drs Ya. Gutak ("Zapsibgeologiya", Novokuznetsk) and E. Yolkin (Institute of Petroleum Geology, Novosibirsk) who donated the material and provided the stratigraphic information. A. Ivanov acknowledges the Sepkoski-PalSIRP grant, and O. Rodina's work is supported by RFFR grant N 02-05-64993.

References

Bonaparte, C.L. 1838: Synopsis vertebratorum systematis. *Nuovi Annali delle Scienze Naturali, Bologna 11*, 105–133.

Denison, R.H. 1979: *Acanthodii. In* Schultze, H.-P. (ed.): *Handbook of Paleoichthyology*, Vol. 5. Gustav Fischer, Stuttgart.

Ginter, M. & Ivanov, A. 1992: Devonian phoebodont shark teeth. *Acta Palaeontologica Polonica 37*, 55–75.

Ginter, M. & Ivanov, A. 1996: Relationships of *Phoebodus. Modern Geology 20*, 263–274.

Ginter, M. & Ivanov, A. 2000: Stratigraphic distribution of chondrichthyans in the Devonian on the East European Platform margins. *Courier Forschungs-Institut Senckenberg 223*, 325–339.

Gutak, Ya. M., Lyakhnitskiy, V.N., Rodygin, S.A. & Fedak, S.I. 2001. Conodonts in the Middle and Upper Devonian sections of south-eastern part of Gorniy Altay. *In* Korobeynikov, A.F. (ed.):

Regional'naya Geologiya. Geologiya mestorozhdeniy poleznykh iskopaemykh. [Records of the International Scientific Conference "Mining and Geological Education in Siberia. 100 Years in the Science and Production Service"], 44-49, Tomsk (in Russian).

Hampe, O. 1989: Revision der *Triodus*-Arten (Chondrichthyes: Xenacanthida) aus dem saarpfälzischen Rotliegenden (Oberkarbon – Perm, SW-Deutschland) aufgrund ihrer Bezahnung. *Paläontologische Zeitschrift 63,* 79–101.

Hampe, O. & Aboussalam, Z.S. 2002: Evidence of a new species of *Omalodus* (Elasmobranchii: Omalodontida) from the Middle Devonian of the Northern Gondwana margin in Morocco. *Journal of Vertebrate Paleontology 22 (Suppl.),* 62.

Hampe, O. & Long, J.A. 1999: The histology of Middle Devonian chondrichthyan teeth from southern Victoria Land, Antarctica. *Records of the Western Australian Museum, Supplement 57,* 23–36.

Huxley, T.H. 1880: On the application of the laws of evolution to the arrangement of the Vertebrata and more particularly the Mammalia. *Proceedings of the Zoological Society of London 1880,* 649–662.

Ivanov, A. & Derycke, C. 1999: Distribution of the Givetian *Omalodus* shark assemblage. Abstracts of the Meeting of IGCP 406 project "Lower–Middle Palaeozoic Events Across the Circum-Arctic", Jurmala, Latvia, September–October 1999. *Ichthyolith Issues, Special Publication 5,* 22–24.

Ivanov, A. & Rodina, O. 2001: Middle and Late Devonian vertebrate biostratigraphy of SW Siberia, Russia. Abstracts of the 15th International Senckenberg Conference, Frankfurt am Main, Germany, May 2001, 50.

Ivanov, A., Vyushkova, L. & Esin, D. 1992: Ichthyofauna. *In* Rzhonsnitskaya, M.A. *et al.* (comp.): *Key Sections of the Middle/Upper Devonian and Frasnian/Famennian Boundary Beds in the Marginal Part of Kuznetsk Basin. Guidebook of a Field Trip of the Devonian Commission,* 89–91, Novosibirsk (in Russian).

Janvier, P. 1977: Les poissons dévoniens de l'Iran central et de l'Afghanistan. *Mémoires de la Société Géologique du France 8,* 277–289.

Long, J.A. & Young, G.C. 1995: Sharks from the Middle–Late Devonian Aztec Siltstone, southern Victoria Land, Antarctica. *Records of the Western Australian Museum 17,* 287–308.

Miller, R.F., Cloutier, R. & Turner, S. 2003: The oldest articulated chondrichthyan from the Early Devonian period. *Nature 425,* 501–504.

Traquair, R.H. 1893: Notes on the Devonian fishes of Campbellton and Scaumenac Bay in Canada. *Geological Magazine N.S. Dec. III, 10,* 145–149.

Turner, S. 1995: *Doliodus problematicus* (Woodward 1892) from the Lower Devonian of Campbellton, northern New Brunswick, the oldest non-marine xenacanthoid? *Ichthyolith Issues 16,* 39–40.

Turner, S. 1997: "*Dittodus*" species of Eastman 1899 and Hussakof and Bryant 1918 (Mid to Late Devonian). *Modern Geology 21,* 87–119.

Turner, S. 1998: The Omalodontida and the appearance of teeth in "sharks". *Ichthyolith Issues, Special Publication 4,* 51–52.

Turner, S. & Campbell, R. 1993: *Doliodus problematicus* (Woodward 1892) from the Lower Devonian of Campbellton, northern New Brunswick, the oldest non-marine xenacanthoid? Abstracts of the Gross Symposium, Göttingen, 4–6 August 1993.

Woodward, A.S. 1892: On the Lower Devonian fish-fauna of Campbellton, New Brunswick. *Geological Magazine N.S. Dec. III, 9,* 1–6.

Yolkin, E.A., Gratsianova, R.T., Iziokh, N.G., Yazikov, A.Yu & Bakharev, N.K. 1997: Devonian sea-level fluctuations on the south-western margin of the Siberian continent. *Courier Forschungs-Institut Senckenberg 199,* 83–98.

Powichthys spitsbergensis sp. nov., a new member of the Dipnomorpha (Sarcopterygii, lobe-finned fishes) from the Lower Devonian of Spitsbergen, with remarks on basal dipnomorph anatomy

GAËL CLÉMENT & PHILIPPE JANVIER

Clément, G. & Janvier, P. **2004 06 01**. *Powichthys spitsbergensis* sp. nov., a new member of the Dipnomorpha (Sarcopterygii, lobe-finned fishes) from the Lower Devonian of Spitsbergen, with remarks on basal dipnomorph anatomy. *Fossils and Strata*, No. 50, pp. 92–112. France. ISSN 0300-9491.

The Powichthyidae (Dipnomorpha, Sarcopterygii) were hitherto known from a single species, *Powichthys thorsteinssoni*, from the Lochkovian–Pragian of Arctic Canada. New material from the Early Devonian of Spitsbergen is referred here to a new species *Powichthys spitsbergensis* sp. nov., and provides additional data on the skull, dermal palate, palatoquadrate, post-orbital, and scales of the Powichthyidae. In particular, it reveals new morphological features, such as an internal process of the lacrimal, a strong anteroventral process of the scales, and the presence of large, upper oral dental plates, which cover the palate and suggest a crushing palatal bite, as in lungfishes. The structure of the skull agrees with Jessen's description of *Powichthys thorsteinssoni*, except that the junction between the supra-orbital and infra-orbital sensory-line canals is lacking in the Spitsbergen species. The exceptional preservation of the dermal elements of the palate shows that the presumed choana is in fact absent in the Powichthyidae, as in the Porolepiformes. The shape of the nasal capsule and the size and arrangement of the adjoining nerve canals are similar to those of the Porolepiformes, but contrary to Jessen's assumption, the palatoquadrate seems to be significantly different.

Key words: Sarcopterygii; Dipnomorpha; Lower Devonian; Spitsbergen; new species *Powichthys spitsbergensis.*

G. Clément [gael.clement@wanadoo.fr] & P. Janvier*[janvier@mnhn.fr], UMR 8569, Département Histoire de la Terre, Muséum National d'Histoire Naturelle, 8 rue Buffon, 75005 Paris, France [*also Honorary Research Fellow, Palaeontology Department, The Natural History Museum, Cromwell Road, London SW7 5BD, UK]

Abbreviations used in figures

an.na, anterior nostril; ant.pr, anterior process of the *pars metapterygoidea* of the palatoquadrate; ant.Te, anterior tectal; ap.pr, apical process of the *pars autopalatina* of the palatoquadrate; art.e, area of articulation between the ethmoid and ethmoidal process of the palatoquadrate; art.m.eth, medial ethmoidal articular area of the *pars autopalatina* of the palatoquadrate; av.pr, anteroventral process of scales; b.f, buccohypophysial foramen; bas.pr, basicranial process; bp.pr, basipterygoid process; com.la, commissural lamina of the palatoquadrate; Der, dermopalatine; Der.Ect, dermopalatine-ectopterygoid; dlpr, dorsolateral process of the *pars autopalatina* of the palatoquadrate; Ept, entopterygoid; eth.com, ethmoidal commissure; fe.vl, *fenestra ventrolateralis*; fo.Der, fossa for dermopalatine fang; fo.not, notochord fossa; hy, articular area for the hyomandibular; hyp.fo, hypophysial fossa; i.m.sc, depression for the insertion of the subcranial muscles; iLa.pr, internal process of the lacrimal; in.art, intracranial articulation; in.ca, internasal cavity; in.cr, internasal crest; in.trans.c, *internasalis transversus* nerve canal; Int1-3, intertemporals 1-3; ioc, infra-orbital sensory-line canal; La, lacrimal; leb.mi, depression for the insertion of the *levator bulbi* muscle; lopc, lateral profundus nerve canal; Max, maxilla; me.pr, median process of the *pars metapterygoidea* of the palatoquadrate; mopc, medial profundus nerve canal; Na, nasal; not, notochord canal; o.m, orbital margin; occ, occipital region; olfc, olfactory nerve canal; orc, orbitonasal nerve canal; ov.Int1, area overlapped by the intertemporal 1; ov.Ju, area overlapped by the jugal; ov.La, area overlapped by the lacrimal; ov.Pa-eth, area overlapped by the parieto-ethmoidal shield; ov.Po, area overlapped by the post-orbital; ov.Pre, area overlapped by the pre-spiracular; ov.Sq, area overlapped by the squamosal; Pa, parietal; pa.au, *pars autopalatina* of the palatoquadrate; Pa.pi, parietal pit-line; pa.pt, *pars metapterygoidea* of the palatoquadrate; Pa-eth, parieto-ethmoidal shield; Par, parasphenoid; pat.pr?, *processus paratemporalis?*; pf.pq, platform of the palatoquadrate; pin.f, pineal foramen; pin.pl, pineal plate; Pmx, premaxilla; Po, post-orbital; post.na, posterior nostril; post.Te, posterior tectal; Ppa, post-parietal; Ppa.pi, post-parietal pit-lines; pq, palatoquadrate; pr.d.eth, descending process of the sphenoid; Pro, post-rostral mosaic; "ps.dp", "parasphenotic dental plate"; soc, supra-orbital sensory-line canal; soc.p, pores of supra-orbital sensory-line canal; spir.m, spiracular margin; Su1-2, supratemporals 1-2; Ta, tabular; Ta.pi, tabular pit-line; uodp, upper oral dental plates; Vo, vomer.

Introduction

A specimen showing the anterior division of the skull of a sarcopterygian from the Wood Bay Formation (Lower Devonian) of Spitsbergen, found during the 1969 French expedition, was passed on to one of us (GC) by Dr Daniel Goujet [Muséum National d'Histoire Naturelle (MNHN), Paris] in late 2000. This specimen (MNHN SVD 2156) showed an open pineal foramen and independent premaxillae, and appeared thus to be different from the classical *Porolepis* snouts from the Wood Bay Formation. The surrounding matrix was a fine-grained, slightly calcareous sandstone, and could be processed by formic acid preparation, completed by mechanical needle preparation. After time-consuming preparation, the specimen revealed the braincase, numerous articulated dermal elements of the palate, parts of the two palatoquadrates, the left post-orbital, and various scales. Such preservation, without any major deformation, is exceptional for a Devonian sarcopterygian. A second, smaller anterior division of a skull (MNHN SVD 2059) also turned up in the 1969 Spitsbergen collection, and also shows a pineal foramen and independent premaxillae that are not traversed by the infra-orbital sensory-line canal. However, its matrix consists of coarse sandstone, so preparation of the ventral surface of this very fragile specimen could not been undertaken.

These two specimens are here considered to belong to a new species of the genus *Powichthys*, a taxon currently interpreted either as the sister group of all other Dipnoiformes [i.e. *Powichthys* (*Youngolepis* (*Diabolepis* (Dipnoi))); Cloutier & Ahlberg 1996], or placed in a trichotomy with *Youngolepis* and the clade including *Diabolepis* and the Dipnoi (Ahlberg & Johanson 1998).

The material described below is held in the collection of the Laboratoire de Paléontologie, MNHN, Paris. Other material mentioned in the text is held in the following institutions: Georg-August University of Göttingen, Germany (Gö), Sedgwick Museum, Cambridge, UK (SMC), and Geological Museum, University of Copenhagen, Denmark (MGUH).

Geological setting and biostratigraphic correlation

The two specimens described herein come from the outcrops of Sigurdfjellet (Mount Sigurd) and Kronprinshøgda in the northeastern part of Haakon VII land, northern Spitsbergen (Vestspitsbergen). The Lower Devonian red sandstones exposed at these two localities belong to the Sigurdfjellet Faunal Division; that is, the lowermost part of the Wood Bay Formation (Goujet 1984; Blieck *et al.* 1987), and are regarded as late Lochkovian to early Pragian in age.

The genus *Powichthys* was previously known from a single species, *Powichthys thorsteinssoni* Jessen, 1975, from the marine limestone of the Drake Bay Formation, on the west coast of Prince of Wales Island, Arctic Canada (Jessen 1975, 1980). The associated fauna of conodonts, brachiopods, trilobites (see references in Jessen 1980) and diverse vertebrates, including thelodonts, heterostracans, placoderms, chondrichthyans, and acanthodians (Vieth 1980), suggest that the upper assemblage of the Drake Bay Formation is much the same age as the Lochkovian Ben Nevis Formation of Spitsbergen. However, the Upper Member of the Peel Sound Formation, which is the contemporary of the Upper Member of the Drake Bay Formation (Elliott 1984), may be referred to the upper Lochkovian and lower–middle Pragian. This age assignment is supported by the presence of the thelodont *Turinia pagei*, the acanthodians *Gomphonchus sandelensis* and *Nostolepis*, and the conodont *Pelekysgnathus serratus serratus* (Vieth 1980). Recent works (Märss *et al.* 1998; Elliott *et al.* 1998) support the correlation proposed by Elliott (1984), Blieck (1984) and Blieck *et al.* (1987) between the Arctic localities (e.g. Prince of Wales Island) and the upper part of the Red Bay Group (Ben Nevis Formation) of Spitsbergen. In contrast, no recent work seems to confirm the correlation between the upper part of the Drake Bay Formation and the Sigurdfjellet Faunal Division (Elliott 1984; Blieck & Cloutier 2000).

The new *Powichthys* species from Spitsbergen described below is thus most probably early Pragian in age, and the Canadian species, *Powichthys thorsteinssoni*, is most probably either late Lochkovian or early Pragian.

According to various authors (e.g. Thorsteinsson 1967; Elliott 1984; Elliott & Dineley 1991; Elliott *et al.* 1998; Blieck & Cloutier 2000), the Lower Devonian fauna of Prince of Wales Island seems to show a marked endemism, and this Arctic region has also been identified as a centre of origin and adaptive radiation for certain heterostracans (Thorsteinsson 1967; Elliott 1984; Elliott *et al.* 1998), particularly the pteraspidids (Elliott & Dineley 1983; Pernegre, this issue). According to current palaeogeographical data, what is now the Prince of Wales and Spitsbergen islands belonged to closely situated areas in the Early Devonian (Dineley & Loeffler 1993; Young 1993). They represent two large, neighbouring sedimentary basins, the Franklinian Basin (north Laurentia) and Caledonian Basin (northwestern Baltica), respectively, both located in the north of the Old Red Sandstone Continent (Janvier & Blieck 1993; Blieck & Cloutier 2000).

Systematic palaeontology

Subclass Sarcopterygii Romer, 1955

Superdivision Dipnomorpha *sensu* Ahlberg, 1991

Family Powichthyidae Jessen, 1980

Remarks. – Jessen (1980) placed the type species *Powichthys thorsteinssoni* in a new family, Powichthyiidae. Following the International Code of Zoological Nomenclature, the correct spelling is Powichthyidae.

Genus *Powichthys* Jessen, 1975

Type species. – *Powichthys thorsteinssoni* Jessen, 1975.

Emended diagnosis. – Dipnomorpha with an anterior (ethmosphenoid) division of the skull roof about 1.5 times longer than the posterior (otoccipital) division; longitudinal row of bones lateral to the post-parietal and parietal, the boundary separating the ethmosphenoid and otoccipital shields being transversely incomplete; post-parietal and supratemporal not fused; transverse pit-lines of post-parietal and tabular in straight continuation; independent premaxilla, with the infra-orbital sensoryline canal passing dorsally (in the suture with the parieto-ethmoidal shield) and not through the bone; small orbital opening; large upper oral dental plates; well-developed internasal cavities; large and well-defined area of the ethmoid for the ethmoidal articular process of the palatoquadrate; presence of a descending process of the basisphenoid; three pairs of suprachordal arcual plates, closely sutured together; no suprapterygoid process; profundus and trigeminus nerves emerging through foramina in the ethmosphenoid, and foramen for the lateral ophthalmic nerve branch located in the otoccipital.

Remarks. – The dermal elements doubtfully referred to by Jessen (1980) as a "Porolepiformes gen. et sp. indet." (i.e. a palatoquadrate complex, a lacrimal, a lateral extrascapular, a lower jaw, bones of the submandibular and operculogular series, a cleithrum, and scales) were considered by Ahlberg (1991, p. 243) to belong to the genus *Powichthys*, because "all the material comes from a single locality, and there is no positive evidence that more than one genus is represented".

Powichthys spitsbergensis sp. nov.
Figs. 1A, B, 2, 3, 4A1–A3, 5, 6B, 7–11, 12A–C

Etymology. – From "Spitsberg", the French translation of Spitsbergen, main island of the Svalbard archipelago, where the specimens were discovered.

Diagnosis. – A *Powichthys* with no pit in the roof of the internasal cavity, and lacking the median, posteriorly directed palatal process of the premaxilla, and the distinct, small teeth in front of the internasal cavity on the narrower portion of the palatal lamina. Parasphenoid narrow, and bearing very few teeth; hypophysial fossa large and apparent. Scales with large, anteroventral processes.

Holotype. – Anterior part of the skull with articulated palatal elements, left post-orbital and scales (MNHN SVD 2156). Collection of Laboratoire de Paléontologie, MNHN, Paris, France.

Referred specimen. – Anterior part of the skull (MNHN SVD 2059). Collection of the Laboratoire de Paléontologie, MNHN, Paris, France.

Localities and horizon. – Kronprinshøgda (MNHN SVD 2156), beneath Karlsbreen glacier, north of Risefjella (type locality), and Sigurdfjellet (MNHN SVD 2059), both in Haakon VII land, northern Spitsbergen, and from the Sigurdfjellet Faunal Division, lowermost Wood Bay Formation, Lower Devonian (late Lochkovian–early Pragian).

Description. –

Parieto-ethmoidal shield
The snout (Figs. 1A, B, 4A) is short and broad relative to that of the Porolepiformes. In specimen MNHN SVD 2156 (Fig. 4A), the sutures between the bones are conspicuous, especially in the posterior region of the shield. The parietals, the posterior part of the post-rostral mosaic, and the posterior nasals are clearly delimited (Pa, Pro, Na, Fig. 4A). The posterior margins of the parietals are damaged and the intertemporals are missing; it is thus impossible to provide a precise description of the structure of the dermal articulation (or suture) between the anterior and posterior divisions of the skull roof. In *Powichthys thorsteinssoni*, and contrary to the Porolepiformes, the transverse suture between the two divisions is bounded laterally by a posterior intertemporal (Int3, Fig. 4B; "dermosphenotic" in Jessen 1975, 1980), thus preventing any mobility on the intracranial joint of the braincase.

The premaxillae (Pmx, Figs. 1B, 2, 4A) are independent from the parieto-ethmoidal shield, and are not traversed by the infra-orbital sensory-line canal (ioc, Figs. 1B, 4A, 5), which instead passes dorsally, within their suture with the parieto-ethmoidal shield. This condition is known in other Dipnomorpha: *Youngolepis* (Fig. 1C; Chang 1982, 1991), *Diabolepis* (Chang & Yu 1984; Chang 1995), and *Powichthys* (Jessen 1975, 1980), as well as some early "osteolepidids" (Fig. 1D, E; Chang & Yu 1997). A

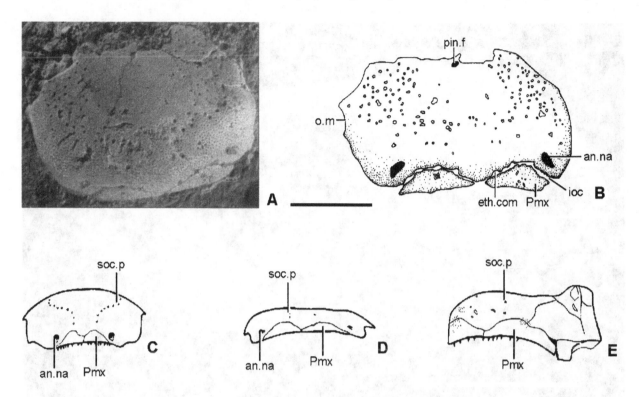

Fig. 1. A, B: *Powichthys spitsbergensis* sp. nov., Wood Bay Formation, Lower Devonian, northern Spitsbergen. Anterior part of the parieto-ethmoidal shield (MNHN SVD 2059). A: Dorsal view. B: Anterior view. Scale bar = 10 mm. C–E: Parieto-ethmoidal shields of other sarcopterygians showing independent premaxillae, in anterior view. Not to scale. C: *Youngolepis praecursor* Chang & Yu, 1981. D: *Thursius wudingensis* Fan, 1992. E: *Kenichthys campbelli* Chang & Zhu, 1993.

specimen of *Porolepis* (SMC J554) also shows independent premaxillae (Jarvik 1972, pl. 4:1; Ahlberg 1991, fig. 5), but the infra-orbital sensory-line canal runs across their dorsal process. Contrary to *Porolepis* (Ahlberg 1991, fig. 5), a dorsal process is lacking on the premaxilla of *Powichthys*.

The infra-orbital sensory-line canal is situated in the lateral part of the posterior suture of the premaxillae, and the ethmoidal commissural sensory-line canal lies in the medial part of the posterior suture of the bone (eth.com, Figs. 1B, 4A, 5). This course of the sensory-line canal suggests that the premaxilla is a single dermal component, and not a composite bone formed by the fusion of the rostral or nasal bones as interpreted by Jessen (1980). A canal-bearing premaxilla has been proposed as a primitive osteichthyan character (Ahlberg 1991). According to Yu (1998, fig. 1B), *Psarolepis* shows an infra-orbital sensory-line canal running through the premaxilla, but its anterior course remains unknown.

The premaxillae of specimen MNHN SVD 2156 bear some broken teeth, with the histological structure of polyplocodont-type, as in the Tetrapodomorpha, *Psarolepis* and *Youngolepis*, and no trend is shown towards the extremely complex dendrodont folding of porolepiform teeth. Vorobyeva (1977) made a detailed study of the teeth of *Powichthys thorsteinssoni*, and considered them to show an eusthenodont-type of folding, as in the teeth of the Tristichopteridae; i.e. the bone of attachment penetrates between the primary folds of the dentine, and the pulp cavity is filled with either bone or osteodentine. Contrary to *Powichthys thorsteinssoni* (Jessen 1980, fig. 4), *Powichthys spitsbergensis* shows no additional teeth in front of the premaxillary teeth. The premaxillary teeth of *Powichthys* are large and well spaced, and situated either near the premaxillary symphysis, or anterior to the internasal cavity, on the narrower portion of the palatal lamina. *Powichthys spitsbergensis* has about 10 teeth on each premaxilla, whereas the "Porolepididae" possess about 20 sharp and closely set teeth on each premaxilla. Contrary to *Powichthys thorsteinssoni*, *Powichthys spitsbergensis* clearly lacks a median, posteriorly directed palatal process of the premaxilla. The rostral region, which extends posteriorly to the premaxillary symphysis, is slightly eroded and no rostral tubule is visible. Rostral tubules are present in dipnoans, *Diabolepis*, and *Youngolepis*.

The anterior external nostril is drop-shaped in anterior view, with a ventrolaterally directed posterior corner (an.na, Figs. 1B, 4A, 5). In both *Powichthys* species this opening is oblique relative to the premaxillary edge, as in *Youngolepis* (an.na, Fig. 1C; Chang 1982, figs. 5D, 6C, pls 6:2, 7:2), whereas it is oval in shape and more or less parallel to the premaxillary edge in the Porolepiformes, like the unique external nostril of the "Osteolepiformes".

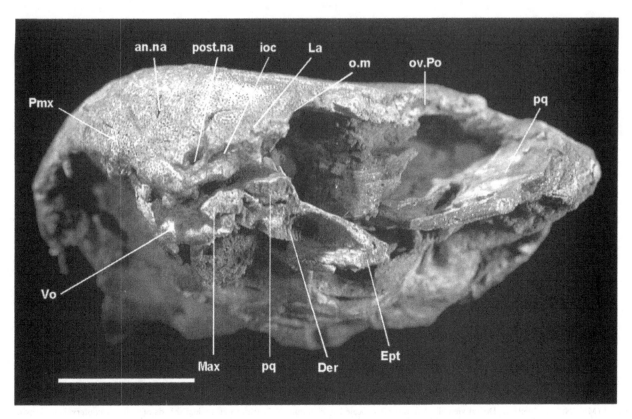

Fig. 2. *Powichthys spitsbergensis* sp. nov., Wood Bay Formation, Lower Devonian, northern Spitsbergen. Anterior division of a skull in left anterolateral view (holotype, MNHN SVD 2156a). Scale bar = 10 mm.

In external view, a *processus dermintermedius* is discern-ible on the internal wall of the anteroventral part of the narinal margin. Jessen (1975, 1980) referred to this pro-cess as the *crista rostrocaudalis* (Jessen 1980, pl. 5:3, "cr.rc") and the *crista rostrocaudalis* proper as a "ridgelike incrassation" (Jessen 1980, pl. 5:3–4, "ri"). A *processus dermintermedius* was considered to be present in *Powichthys* by Zhu *et al.* (2001) and Zhu & Schultze (2001), yet without any detailed argument. The presence of an internal process of the tectal cannot be ascertained in the *Powichthys* specimens known to date. This process is present in the Tetrapodomorpha and *Youngolepis* (Chang 1982, fig. 3).

The anterior part of the left lacrimal, and a small part of the anterior region of the left maxilla, are preserved in the holotype of *Powichthys spitsbergensis*, in connection with the parieto-ethmoidal shield (La, Max, Figs. 2, 5, 8). Although the lacrimal is very slightly displaced postero-dorsally, the relationships between the bones surround-ing the posterior external nostril are quite clear. As in *Porolepis*, the posterior external nostril (post.na, Figs. 2, 3, 5) is delimited anteriorly by a notch in the anterolateral margin of the parieto-ethmoidal shield, and situated dor-sally to the posterior end of the premaxilla. The posterior external nostril is also delimited posteriorly by a notch in the anterior margin of the lacrimal. In *Youngolepis*, the anterolateral margin of the parieto-ethmoidal shield and

the dorsal margin of the posterior end of the premaxilla delimit the anterior part of the posterior external nostril (Chang 1982) in approximately the same way as in *Powichthys* and *Porolepis*. However, the notch supposed to be present in the anterior margin of the lacrimal of *Youngolepis* is not visible in the specimen figured and described by Chang (1991, fig. 5, pl. 1A, C). In *Powichthys spitsbergensis*, the posterior external nostril is oval in shape (post.na, Figs. 2, 3, 5). The anteroventrolateral corner of the parieto-ethmoidal shield lies very close to the anteroventral end of the lacrimal, and thus bounds the ventral border of the posterior external nostril. A very small gap may be present between these two bones, and it is thus possible that the posterior end of the premaxilla contributed to the ventral border of the posterior external nostril as well. The infra-orbital sensory-line canal (ioc, Figs. 2, 3, 5), which runs in the suture between the parieto-ethmoidal shield and the posterior part of the premaxilla, continues backwards within the anter-oventral expansion of the lacrimal. The foramen for this canal is situated ventrally to the notch that forms the posterior margin of the posterior external nostril. Inside the lacrimal, the infra-orbital sensory-line canal runs posterodorsally towards the orbit.

An important feature is the presence of a strong internal process of the lacrimal (iLa.pr, Fig. 3). This process, which is clearly visible on the internal margin of

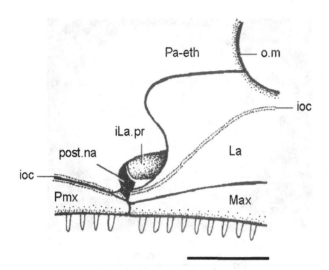

Fig. 3. *Powichthys spitsbergensis* sp. nov., Wood Bay Formation, Lower Devonian, northern Spitsbergen. Restoration of the posterior nostril of the left side, with adjacent bones, based on the holotype. Scale bar = 5 mm.

the posterior external nostril, is slightly concave dorsally and projected medially. Its anterior end is situated very close to the posterior end of the *crista rostrocaudalis*, and its anterior part extends backwards to the orbit, but its posterior extension cannot be determined. Such a process has never been recorded previously, and is also present on the lacrimal of two specimens of *Porolepis* (MNHN SVD 2038, 2159), figured by Jarvik (1972, pls 7:3–4; 6:1) where it contacts with the parieto-ethmoidal shield. This process is absent in the Holoptychiidae, as shown by grinding series (sections 62, 67) of the anterior region of the skull of *Glyptolepis groenlandica* (Jarvik 1972), and preserved lacrimals of other Holoptychiidae (Jarvik 1972, pls 16:1, 24:1, 24:4, 29:3, 29:5). The lacrimal of *Powichthys thorsteinssoni* as described and figured by Jessen (1980, pl. 4:2) has a very similar anterior part to that of *Powichthys spitsbergensis*, but the internal surface has not been prepared, and it remains unknown whether an internal process is present or not.

Powichthys spitsbergensis (and *Porolepis*) thus provide the first evidence for a strong internal process of the lacrimal in sarcopterygians. As currently known, this feature is shared only by *Powichthys* and *Porolepis*, but it is worth noticing that the lacrimal of the holoptychiids *Quebecius quebecensis* (Cloutier & Schultze 1996, fig. 14A) and *Glyptolepis groenlandica* (specimen MGUH P.1510; Jarvik 1972, figs. 7B, 79D, pl. 17:1–3) shows a well-marked groove running from the notch forming the posterior margin of the posterior external nostril to the anteroventral margin of the orbit. Jarvik (1972) assumed that this groove housed an external nasolacrimal duct, but the presence of a nasolacrimal duct has been regarded as a tetrapod character. The curved shape and position of the internal process of the lacrimal of *Powichthys* and

Porolepis, and the position of the internal groove of the lacrimal in *Quebecius* and *Glyptolepis groenlandica*, both strongly suggest a canal leading from the nasal capsule to the orbit. As these two structures (internal process and groove of the lacrimal) are in the same position, they are likely to be linked with the same, homologous structure of the soft anatomy.

The pineal foramen of *Powichthys spitsbergensis* is conspicuous in both specimens (pin.f, Figs. 1B, 4A). It is not situated at the top of a bulge, like that often described in the anterior part of the parieto-ethmoidal shield of the Porolepiformes, *Youngolepis* and *Diabolepis*. In fact, the "pineal" bulge that may sometimes be observed in the latter taxa is likely to be an artefact of post-mortem deformation due to the presence of a more densely ossified region in the underlying neurocranium, at the junction between the two canals for the olfactory tracts. Contrary to the condition in "Osteolepiformes", the pineal foramen of *Powichthys* is not situated between the parietals, but more anteriorly. In *Powichthys thorsteinssoni* it opened through a dermal pineal plate (pin.pl, Fig. 4B), but in the holotype of *Powichthys spitsbergensis* it is placed in a suture of the posterior region of the post-rostral mosaic (Pro, Fig. 4A). It is impossible to decide whether the pineal canal is single or paired. In the Porolepiformes there is a pineal canal *sensu stricto* and a parapineal canal (Jarvik 1972; Bjerring 1975).

Although the posterior part of the parietals and intertemporals are missing, as far as preserved the bone pattern of the parieto-ethmoidal shield agrees with that of *Powichthys thorsteinssoni*, apart from differences in the number and proportions of the bones of the post-rostral mosaic. As Cloutier (1996, 1997) noted for the Dipnoi, the Dipnomorpha generally display significant intraspecific variation in the skull roof pattern.

The supra-orbital sensory-line canal (soc, Fig. 4A) runs posterolaterally and is S-shaped. It passes through the nasal series and continues posteriorly either inside the anterior part of the parietal, or along the suture between the parietal and the laterally adjacent bones.

A very large overlap area for the anterior intertemporal (ov.Int1, Fig. 4A) is present laterally to the left parietal. Such an overlap area is unknown in other sarcopterygians, and is not clearly visible in *Powichthys thorsteinssoni*. In dorsal view, this very large area is laterally concave and medially convex. Medially, a well-marked groove runs along the lateral margin of the parietal. This groove is continued anteriorly by a canal situated in the suture between the parietal and the posterior nasal. When the anterior intertemporal was articulated to the parietal, this groove certainly formed the floor of the posterior portion of the supra-orbital sensory-line canal (soc, Fig. 4A).

The lateral margin of the overlap area for the anterior intertemporal is thickened and displays anteriorly a well-marked notch (white arrow, Fig. 4A2). The left

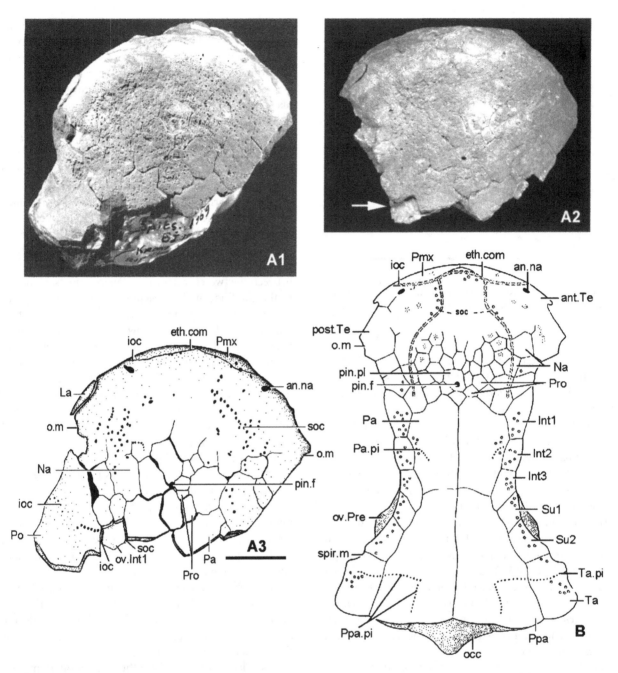

Fig. 4. A: *Powichthys spitsbergensis* sp. nov., Wood Bay Formation, Lower Devonian, northern Spitsbergen. Anterior part of the parieto-ethmoidal shield in dorsal view (holotype, MNHN SVD 2156a). A1: Before preparation with left post-orbital *in situ*. A2: After preparation, without left post-orbital (white arrow: notch in the anterior part of the lateral margin of the overlap area for the anterior intertemporal). A3: A drawing of the prepared specimen with the left post-orbital in connection. Scale bar = 10 mm. B: *Powichthys thorsteinssoni* Jessen, 1975. Skull roof in dorsal view (from Jessen 1975). Not to scale.

post-orbital (Po, Figs. 4A, 11), found in connection with the skull, shows a dorsal foramen for the infra-orbital sensory-line canal situated opposite the notch just mentioned. The groove which passes by the notch, i.e. the floor of the infra-orbital sensory-line canal (ioc, Fig. 4A) when the anterior intertemporal was articulated, bends posteriorly at a right angle. The infra-orbital sensory-line canal thus follows the thick margin of the overlap area without any junction with the supra-orbital sensory-line

canal. This demonstrates that there is no junction between the infra- and supra-orbital sensory-line canals, contrary to Jessen's (1980) assumption, and in accordance with Ahlberg's (1991) hypothesis. This junction is also lacking in *Youngolepis* (Chang 1982), *Diabolepis* (Chang 1995), and the Dipnoi (Campbell & Barwick 1987), as well as in actinistians and actinopterygians. This condition is considered as primitive, or general, for osteichthyans.

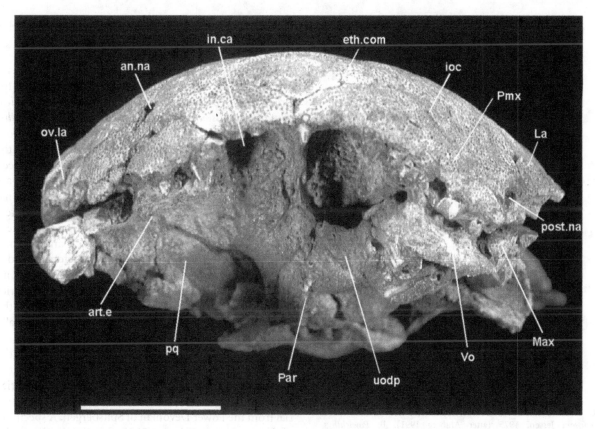

Fig. 5. *Powichthys spitsbergensis* sp. nov., Wood Bay Formation, Lower Devonian, northern Spitsbergen. Anterior division of a skull with associated dermal elements of the buccal roof in anterior view (holotype, MNHN SVD 2156a). Scale bar = 10 mm.

Ethmoidal region of the ethmosphenoid

The ethmoidal region of *Powichthys spitsbergensis* is most similar to that of *Powichthys thorsteinssoni* and the Porolepiformes. The internasal cavities are large and deep (in.ca, Figs. 5, 7, 8), and separated by an internasal crest. Considering their shape and the presence of an area for a parasymphysial dental plate at the anterior end of the lower jaw of *Powichthys thorsteinssoni* (Jessen 1980, fig. 1B, C), the internasal cavities certainly housed the fangs of the parasymphysial dental plates, as in the Porolepiformes and Onychodontida. However, in *Powichthys*, the floor of these cavities displays a network of interconnected grooves, whereas it is smooth in the Porolepiformes. The internasal crest of *Powichthys* is thinner than in Porolepiformes and also bears some interconnected ridges. *Powichthys* has a "mushroom-like" structure behind the internasal crest; this is less developed in *Powichthys spitsbergensis* than in *Powichthys thorsteinssoni* (Jessen 1980, fig. 4, pl. 5:1, "m.s"). Its shape is more or less globular, but contrary to Jessen's description of *Powichthys thorsteinssoni*, this structure is not pierced by numerous small canals. Its surface displays a network of interconnected, sinuous ridges, but without foramina.

There is a large shallow groove between the *solum nasi* and the anterior part of the parasphenoid but, contrary to the condition in *Powichthys thorsteinssoni*, its anterior end is not a pit in the bottom of the internasal cavity (Jessen 1980, fig. 4, "pi"), but a network of small, interconnected ridges covering the internasal cavity. This groove certainly conveyed vessels and nerves to various soft tissues housed in the internasal cavity.

The nasal cavity proper cannot be observed, but the posterior foramina of the median and lateral profundus canals and the orbitonasal canal, which pierce the post-nasal wall, are conspicuous on the left side of the holotype (MNHN SVD 2156), permitting a tentative reconstruction of the general shape of the nasal capsule (Fig. 6B). The median profundus canal (mopc, Fig. 6B) is large and situated mediolaterally in the post-nasal wall. Five small lateral profundus canals pierce this wall dorsolaterally to the median profundus canal. The orbitonasal canal, situated ventrolaterally to the median profundus foramen, is much smaller than the latter. The position and proportion of these nerve canals are very similar to those in the nasal capsule of *Porolepis* (Fig. 6C) but strikingly different from those in the nasal capsule of *Powichthys thorsteinssoni* (Jessen 1980, pl. 9:3; Ahlberg 1991, fig. 10B). Ahlberg's reconstruction (Fig. 6A) is based on a unique internal mould of the nasal capsule of the skull Gö 100-377 figured by Jessen. This skull is also figured (Jessen

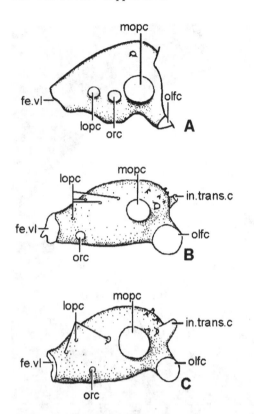

Fig. 6. Nasal capsule of the left side, in posterior view. A: *Powichthys thorsteinssoni* Jessen, 1975 (after Ahlberg 1991). B: *Powichthys spitsbergensis* sp. nov., Wood Bay Formation, Lower Devonian, northern Spitsbergen (holotype, MNHN SVD 2156a). C: *Porolepis* sp. (after Jarvik 1942). Not to scale.

1980, pl. 9:1) and it is obviously much eroded and distorted. The triangular shape of this nasal capsule, compared with the more usual ovoid shape of that of porolepiforms, is most probably due to a post-mortem distortion. Moreover, this internal mould has been naturally exposed to erosion, and this may explain the absence of any trace of the small, lateral profundus canals. The presence of a very large orbitonasal canal (orc, Fig. 6A) close to the median profundus canal (mopc, Fig. 6A) and of a unique large lateral profundus canal (lopc, Fig. 6A) is unusual and rather conjectural. Furthermore, this pattern is not present in the other nasal capsules figured by Jessen (1975, fig. 4A, F, G). It is conspicuous, as previously noticed by Chang & Smith (1992, p. 307), that the latter show an orbitonasal canal situated far from the median profundus canal, and a series of small lateral profundus canals, as in porolepiforms and *Powichthys spitsbergensis*. The internal mould of the nasal capsule of specimen Gö 100-377, used by some authors as reference material for phylogenetic studies, should be left aside or, at least, considered as highly conjectural. Consequently, as Ahlberg (1991, p. 224) put it: "in other sarcopterygians, the two branches [the median profundus canal and the lateral profundus canals] tend to be approximately equal in size. This suggests that a much enlarged median profundus

foramen may be a synapomorphy of *Powichthys* and porolepiforms".

Sphenoidal region of the ethmosphenoid

Some differences between the sphenoidal region of the braincase of *Powichthys spitsbergensis* and that of *Powichthys thorsteinssoni* can be pointed out. In *Powichthys spitsbergensis*, three large and shallow grooves run from the lateral margin of the posteroventral region of the sphenoid and bend towards a median, large and very deep hypophysial fossa (hyp.fo, Fig. 8). From rear to front are seen: the groove for the medial branch of the internal carotid artery, the groove for the main trunk of the internal carotid artery and a groove for the palatine artery running along the posterior part of the parasphenoid. Contrary to the condition in the Porolepiformes and – *Powichthys thorsteinssoni* (Jessen 1980, figs. 4, 5), no separate foramen of the respective canals for these arteries is present in the medial region of the ventral surface of the sphenoid. These three grooves are coalescent medially and bend towards the hypophysial fossa.

Remarkably, the position of the foramina for the branches of the internal carotid arteries and for the palatine artery shows an important intraspecific variation, which has also been noted in new porolepidid material from the Lower Devonian of Spitsbergen. A specimen of *Heimenia* sp. (Clément 2001, fig. 3) also shows the coalescence of the grooves for the two branches of the internal carotid arteries inside a large hypophysial fossa. *Youngolepis* shows a coalescence of the foramina in the ventral surface of the sphenoid (Chang 1982). However, this condition is not the same as in the "Osteolepiformes" and Actinistia, which have only lateral carotid openings, apparently corresponding to the openings for the efferent pseudobranchial arteries in actinopterygians (P. E. Ahlberg 2001, pers. comm.). It should also be noticed that the medial end of the groove for the palatine artery, owing to this coalescence, is situated in the hypophysial fossa, but continues anteriorly along the posterior part of the parasphenoid and bends slightly laterally, piercing the endoskeleton in a very short distance. A foramen for the palatine artery is thus present in the same position as in *Powichthys thorsteinssoni* (Jessen 1980, fig. 5, "c.a.pal"). Another difference between the two species of *Powichthys* is the position of the foramen for the *ophthalmica magna* artery. In *Powichthys thorsteinssoni*, Jessen assumed this was situated anteriorly and ventrally to the basipterygoid process, between the foramen for the internal carotid artery and the foramen for the palatine artery (Jessen 1980, fig. 5, "c.a.om"). In *Powichthys spitsbergensis*, this foramen is in the same position as in *Porolepis* (Jarvik 1972, fig. 20A) and *Youngolepis* (Chang 1982, pl. 15A); that is, dorsal to the anterior end of the basipterygoid process. Its canal passes through the basipterygoid process, and its medial foramen is clearly visible in the lateral wall of the hypophysial fossa.

Fig. 7. Powichthys spitsbergensis sp. nov., Wood Bay Formation, Lower Devonian, northern Spitsbergen. Anterior division of a skull in ventral view, with associated dermal elements of the buccal roof (holotype, MNHN SVD 2156a). A: With the vomer and upper oral dental plates of the right side *in situ*. B: With the vomer and upper oral dental plates of the right side removed. Scale bars = 10 mm.

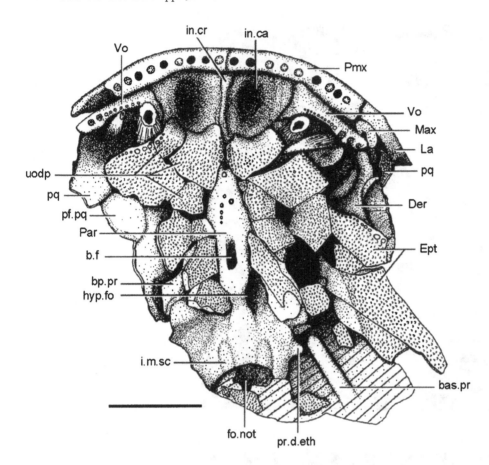

Fig. 8. *Powichthys spitsbergensis* sp. nov., Wood Bay Formation, Lower Devonian, northern Spitsbergen. Anterior division of a skull in ventral view, with associated dermal elements of the buccal roof (holotype, MNHN SVD 2156a). Interpretive drawing of Fig. 7A. Scale bar = 10 mm.

A small foramen situated immediately above the anterior end of the parasphenoid was termed as a "canal probably for vessels" by Jessen (1980, fig. 5, "c.t"). It has the same position in *Powichthys spitsbergensis*, and could be the equivalent of the foramen for the "canal for ascending branch of *r. palatinus* anastomosing with ventral branch of *n. ophtalmicus profundus*" of *Porolepis brevis* (Jarvik 1972, fig. 73, "c.pald"). The buccohypophysial foramen (b.f, Fig. 8) is very large, and pierces the parasphenoid somewhat anteriorly to the hypophysial fossa. A large anteromedial expansion of the sphenoid is fused with a large posteromedial expansion of the parasphenoid, forming a large bridge ventrally to the hypophysial fossa.

The posteroventral region of the sphenoid shows two well-marked depressions (i.m.sc, Fig. 8), which are also clearly visible in *Powichthys thorsteinssoni* (Jessen 1980, figs. 4, 5, "i.m.sc") and in the holotype of *Porolepis brevis* (Jarvik 1972, fig. 65B).These may represent the attachment area for the basicranial muscle (Jessen 1980), but Jarvik placed this attachment area more laterally on the sphenoid in the Porolepiformes (Jarvik 1972, figs. 20A, 21A, pl. 9:2, "i.m.sc").

An interesting feature is the descending process of the basisphenoid (pr.d.eth, Fig. 8), which occurs in *Powichthys thorsteinssoni*, *Youngolepis* (pr.d.eth, Fig. 13A, B) and actinopterygians. In *Powichthys spitsbergensis*, the

left descending process is well developed while that of the right side is vestigial. Both are situated in a more dorsal position than those of *Powichthys thorsteinssoni* and *Youngolepis*, and are lined with perichondral bone. According to Jessen (1980), the descending process of *Powichthys* might have articulated with a subchordal plate. According to Bjerring (1994, fig. 12; 1995, fig. 5), the descending process of the basisphenoid is connected to the basicranial process of the otoccipital by means of a cartilaginous bridge forming the "basicranial ansilla"; i.e. a cartilaginous bridge homologous to the lateral commissure of actinopterygians (Gardiner 1984). A large groove runs anteriorly from the lateral margin of the recess for the anterior end of the notochord (fo.not, Fig. 8), and ends abruptly, dorsal to the descending process of the basisphenoid. The same groove in *Powichthys thorsteinssoni* has been correctly determined as the articular area for the anteroventral process of the otoccipital (Jessen 1980, fig. 5, "art.otoc.v"). The anteroventral process of the left side of the otoccipital (bas.pr, Fig. 8), although broken and slightly displaced, is visible in *Powichthys spitsbergensis*. Its shape, diameter and position confirm that it lies against the groove that Jessen assigned to it, and was not fused to the sphenoid.

This new material of *Powichthys* thus confirms that the descending process of the basisphenoid is not articulated with the anteroventral process of the otoccipital. More-over, it seems that the anterior tip of the anteroventral process of the otoccipital in *Youngolepis* (Chang 1982, pl. 10) is covered with perichondral bone, and thus does not present an attachment surface for any cartilaginous "basicranial ansilla". The presence of this bridge in *Youngolepis*, despite Bjerring's assumption, is thus very doubtful. However, an articulation between the descending process and a subchordal plate cannot be confirmed in *Powichthys spitsbergensis*.

Dermal elements of the mouth cavity

The holotype MNHN SVD 2156 exhibits the vomers, the anterior part of the left dermopalatine, the anteromedial part of the entopterygoid, the parasphenoid, and numerous upper oral dental plates *in situ*. Such a preservation is exceptional for a Devonian dipnomorph, and significantly increases our knowledge of basal dipnomorph anatomy.

The vomers (Vo, Figs. 2, 5, 7, 8) are widely separated from each other and do not contact the parasphenoid. The left vomer has two large fangs, and the right vomer bears one large fang and two smaller ones. Their section shows a structure of polyplocodont-type. The vertical tooth-bearing lamina is higher than that of *Youngolepis* (Vo, Fig. 13B; Chang 1982, fig. 7A) and, as in *Heimenia* (Clément 2001), presents numerous small rounded teeth, randomly arranged anterior to the row of denticles. However, this area covered with small rounded teeth is not expanded into a broad toothed field, as in *Youngolepis* (Vo, Fig. 13B) and *Diabolepis* (Chang 1995, figs. 4, 11). The vertical, tooth-bearing lamina extends further posteriorly than the posterior limit of the base of the vomer, as in *Heimenia* and in the osteolepidid *Medoevia* (Lebedev 1995, fig. 4A). A flat, tongue-like, anteromedial process, which is conspicuous in *Heimenia*, is not present in *Powichthys spitsbergensis*. The medial edge of the vomer is strongly convex and shows an overlap area for the anterolateral upper oral dental plates.

The posterior part of the left dermopalatine is missing, and its anterior part is laterally broken, but its relationships with the anterior adjacent bones can be described (Der, Figs. 2, 7, 8, 10). The anterior tip of the dermopalatine covers the posteromedial edge of the vomer, and not its posterolateral edge, as clearly visible in *Heimenia* (Clément 2001, fig. 5D). The dermopalatine rests anteriorly on the ventral surface of the thick anterior border of the palatoquadrate (Der, Fig. 10) and is, more posteriorly, in firm contact with the ventral side of the very thin pars autopalatina (pa.au, Fig. 10). A large ventral depression housed a dermopalatine fang (fo.Der, Fig. 10). This anterior part of the dermopalatine is bounded posteriorly by the thick, oblique, anterolateral edge of the entopterygoid (Ept, Figs. 2, 8, 10).

The entopterygoid (Ept, Figs. 2, 8, 10) rests firmly on the commissural lamina of the palatoquadrate (com.la, Fig. 10). This lamina is very thin and looks like a leaf covering the dorsal surface of the entopterygoid. The ventral surface of the entopterygoid is covered with numerous, small, rounded denticles, except on its lateral border which has a row of larger teeth (Ept, Fig. 8). The area covered with small rounded denticles is less extended medially and anteriorly than in *Glyptolepis* (Jarvik 1972, fig. 31). The smooth, medial and anterior surface of the entopterygoid is covered with a mosaic of upper oral dental plates (uodp, Fig. 8).

The parasphenoid (Pa, Figs. 5, 8) is elongate in shape, narrow, and tapers anteriorly behind the "mushroom-like" structure of the internasal area of the ethmoid region. Its ventral surface is convex and shows only two folded teeth anteriorly and five small teeth in its middle part. No other tooth or denticle is present on the parasphenoid, whereas they are numerous on the upper oral dental plates and the entopterygoid. The lateral margins of the parasphenoid are overlapped by the adjacent upper oral dental plates. The buccohypophysial foramen (b.f, Fig. 8) is very large, oval in shape and has an oblique anterior wall. The surrounding wall of this canal is covered with small dental plates bearing minute, pointed teeth. Similar denticles occur inside the buccohypophysial canal of *Youngolepis* (Chang 1982, p. 26).

The parasphenoid of *Powichthys spitsbergensis* is very different from that of *Powichthys thorsteinssoni* (Par, Fig. 13A; cf. Jessen 1980, figs. 4, 5, pls 5:1, 6:1–2, 7:3). In the latter, the ventral surface of the parasphenoid is large and flat, and bears numerous small rounded denticles, which become larger anteriorly. The posterior part of the parasphenoid is broad and also covered with numerous denticles, and its posterior end is rounded and situated relatively far posteriorly to the hypophysial fossa. The unusually simple shape of the parasphenoid of *Powichthys spitsbergensis*, and its difference in shape and dentition from that of *Powichthys thorsteinssoni*, confirm Chang's assumption (1982, pp. 27–28), previously alluded to by Jarvik (1954, pp. 65–67) and Jessen (1980, p. 207), that the upper oral dental plates, including the "parasphenotic dental plates" of Jessen and the "prespiracular dental plates" of Chang, became gradually fused to the middle portion of the parasphenoid (termed the "corpus" of the parasphenoid by Chang). According to Chang (1982), a complete absence of fusion is a primitive condition for the dipnomorphs. This primitive condition would be represented in *Powichthys spitsbergensis*, where the parasphenoid (very narrow, convex and almost devoid of teeth) is surrounded by more than 10 upper oral dental plates. The intraspecific variation of the shape of the parasphenoid, as exemplified by *Youngolepis praecursor* (Chang 1982, fig. 8A–C) and *Diabolepis speratus* (Chang & Yu 1984, fig. 2A–D), is explained by differences in the degree of fusion of some originally independent upper oral dental plates with the "corpus" of the parasphenoid.

The upper oral dental plates (uodp, Figs. 5, 7, 8, 9) are very large relative to the numerous, small, upper oral dental plates of the Porolepiformes (Jarvik 1972, figs. 30, 72A). As in the latter, they form a mosaic that covers most of the buccal roof, yet the posterior extension of this mosaic cannot be determined in *Powichthys*. The two upper dental plates in *Powichthys thorsteinssoni* were considered by Jessen (1980, figs. 4, 5) as the parasphenotic dental plates ("ps.dp", Fig. 13A), and show a smooth, concave area on their lateral edges. This area is not the groove for a pre-spiracular gill-pouch (Jessen 1980, figs. 4, 5, "gr.psp"), but an overlap area for the adjacent lateral upper oral dental plates, as can be seen in *Powichthys spitsbergensis*.

The upper oral dental plates of *Powichthys spitsbergensis* are in contact medially with the lateral margins of the parasphenoid, and anterolaterally with the medial margin of the vomers. They rest posterolaterally on the medial part of the entopterygoids and posteriorly on the ventral surface of the sphenoid. The anterior-most plate (or plates; there are two on the right side of MNHN SVD 2156) is situated between the anterior tip of the parasphenoid and the medial margin of the vomer. The plate bends anteriorly at a right angle towards the bottom of the internasal cavity (uodp, Figs. 5, 7, 8, 9A, B). This part is smooth, whereas the posterior part of the plate is covered with numerous, small, rounded denticles. The lateral margin of the anterior plate overlaps the medial margin of the vomer and displays a row of enlarged, rounded teeth (Fig. 9A), which continues backwards and is prolonged by the lateral tooth row of the entopterygoid (Figs. 7, 8). The medial margin of the anterior plate, overlapping the lateral margin of the parasphenoid, also shows a row of enlarged rounded teeth (Figs. 7, 8, 9A).

Except for the enlarged, rounded teeth of the anterior plates, the dentition of the upper oral dental plates consists of numerous, minute, randomly arranged denticles. These are rounded, except on the posterior-most plates and on the plates which are situated along the parasphenoid, where they are pointed, like those of the parotic plates of the osteolepidid *Medoevia* (Lebedev 1995, fig. 15).

The mosaic of upper oral dental plates, the poorly toothed area of the parasphenoid, and the toothed parts of the entopterygoids, thus formed together a large and flat denticulate buccal roof. Such a buccal roof, covered with an extensive, crushing dentition, also occurs in the Dipnoi (e.g. Miles 1977), in *Diabolepis* (Chang & Yu 1984; Chang 1995), most probably in *Youngolepis* (Ahlberg 1991), and the Porolepiformes (Jarvik 1972). The large upper oral dental plates of *Powichthys* may have formed by the fusion of numerous, small upper oral dental plates, as seen in the Porolepiformes. These dental plates are neomorphic elements, functionally correlated with the

Fig. 9. A, B: *Powichthys spitsbergensis* sp. nov., Wood Bay Formation, Lower Devonian, northern Spitsbergen. Anterior upper oral dental plates of the right side (holotype, MNHN SVD 2156a). A: Ventral view. B: Dorsal view. Scale bars = 10 mm.

development of a palatal bite, as in the Dipnoi and *Diabolepis*, yet they cannot be homologues of the palatal tooth plates of the latter two taxa, which are derived from the entopterygoid. In *Powichthys* and the Porolepiformes, the dental plates clearly underlie a large part of the surface of the entopterygoids.

It is worth noting here that some Tristichopteridae (derived "Osteolepiformes") show a small, elongate, dermal element between the parasphenoid, the vomer and the entopterygoid. This element, well visible in *Mandageria* and *Cabonnichthys*, and referred to as the "accessory vomer" (Ahlberg & Johanson 1997; Johanson & Ahlberg 1997), is also regarded as a neomorphic element.

Palatoquadrate

The palatoquadrate of *Powichthys spitsbergensis* (pq, Figs. 2, 5, 7, 8, 10) is made up of a single unit, like that of most basal taxa of the major gnathostome groups, but may be divided arbitrarily into three parts: the *pars autopalatina* (anterior-most part), *pars metapterygoidea* (posterodorsal part) and *pars quadrata* (posteroventral part). The anteromedial region of the palatoquadrate is almost complete in MNHN SVD 2156, missing only its central part. The anterior part of the right palatoquadrate is slightly displaced anteriorly, and the anterior part of the left one is slightly displaced posteroventrally. The *pars metapterygoidea* is present on both sides, but anteriorly displaced.

The *pars autopalatina* (pa.au, Fig. 10) occupies a more or less horizontal position, as in the Middle Devonian "osteolepidids", the Porolepiformes, *Youngolepis*, and *Diabolepis*, all having a platybasic skull, and it closely resembles that of *Eusthenopteron* (Jarvik 1980, fig. 107). As stated by Zhu & Schultze (1997, p. 299): "[Jessen] explained the palatoquadrate of *Powichthys* in a model of porolepiforms. But, as clearly shown by his plates (Jessen 1980, pl. 1:1–2), the ethmoidal articular process and the anteromedial lamina at the anterior end of the *pars autopalatina* identified by Jessen are more suggestive of the apical process of osteolepiforms (Jarvik 1942, 1972; Lebedev 1995)". In *Powichthys spitsbergensis* the apical process (ap.pr, Fig. 10) is well developed, and its anteromedial wall shows a large articular area (art.m.eth, Fig. 10) which is probably homologous to the medial ethmoidal articular area of the palatoquadrate of *Eusthenopteron*. This ethmoidal articular process contacted the large, bean-shaped, articular area of the post-nasal wall. The dorsolateral process of the *pars autopalatina* (dlpr, Fig. 10) is situated laterally and slightly posteriorly to the apical process. A large depression for the insertion of the

levator bulbi muscle of the eye (leb.mi, Fig. 10) is present immediately behind the dorsolateral process. The medial edge of the palatoquadrate, posterior to the apical process, is almost straight. This indicates that, contrary to the condition in the Porolepiformes, an expanded anteromedial articular lamina was lacking, and the palatoquadrate had only one anterior articular process, the apical process, as in the "Osteolepiformes" (Jarvik 1942, 1972, 1980; Lebedev 1995), and not two, as in the Porolepiformes (Jarvik 1972, 1980).

In accordance with Panchen & Smithson (1987), Cloutier & Ahlberg (1996) and Zhu & Schultze (1997), there is no real *fossa autopalatina* in *Powichthys*, i.e. a well-marked depression in the interorbital wall, dorsally to the parasphenoid. However, a shallow depression is present posteromedially to the bean-shaped articular area of the post-nasal wall, and ends medially, dorsally to the posteromedial boundary of the attachment area for the vomer. The anteromedial edge of the palatoquadrate fits in this depression but without any articulation. This shallow depression, also called "*fossa autopalatina*" by Jessen (1980, fig. 4), is different in shape and position from the *fossa autopalatina* of the Porolepiformes (Jarvik 1972) and should preferably not be referred to by the same term. *Youngolepis* (Chang 1982) displays the same condition as in *Powichthys*, i.e. it has no real *fossa autopalatina*, but a shallow depression along the ventral edge of the post-nasal wall for contact with the anteromedial edge of the palatoquadrate. An important difference between the *pars autopalatina* of *Eusthenopteron* and that of *Powichthys spitsbergensis* is that the latter shows a large expansion lateral to the apical process, which is anteriorly thick, dorsally convex, and overlies the dorsal

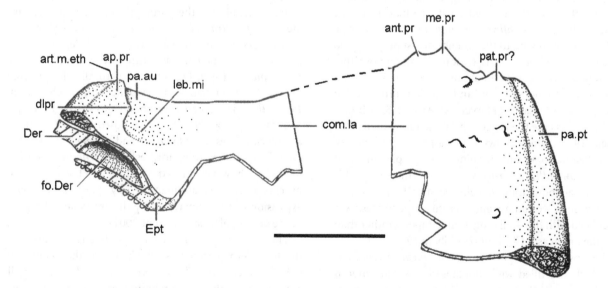

Fig. 10. *Powichthys spitsbergensis* sp. nov., Wood Bay Formation, Lower Devonian, northern Spitsbergen. Anterior and posterodorsal parts of the palatoquadrate and associated dermopalatine and entopterygoid of the left side, in lateral view (based on the holotype, MNHN SVD 2156a). The hatched areas indicate broken surface of bone. The dashed line indicates the reconstructed outline if *processus ascendens* considered absent. Scale bar = 10 mm.

surface of the anterior part of the dermopalatine (Der, Fig. 10).

The limit of the *pars metapterygoidea* (pa.pt, Fig. 10), between the posterodorsal edge and the thick posterior edge that is continued posteroventrally by the *pars quadrata*, is well marked by an almost right angle. This condition is quite different from that in the Porolepiformes and *Eusthenopteron*. The posterodorsal margin of the *pars metapterygoidea* shows a series of small, rounded processes, which are lacking in the Porolepiformes and *Eusthenopteron* but occur in actinopterygians (Arratia & Schultze 1991, fig. 1C, D). The posterior-most of these processes, situated close to the right angle that marks the posteromedial limit of the palatoquadrate, could be considered as the *processus paratemporalis* of "Osteolepiformes" (pat.pr?, Fig. 10). The two, more anteriorly placed processes, referred to here as the "anterior" and "median" processes (ant.pr, me.pr, Fig. 10), are closely set and could be considered as homologues of the *processus ascendens*. In this case, the notch between the *processus paratemporalis* and the "median" process would have given passage to the trigeminal nerve.

According to Chang & Yu (1997): "In forms where the suprapterygoid process is absent or underdeveloped, the ascending process of the palatoquadrate is assumed to be in ligamentous connection with the neurocranium (Jarvik 1972; Vorobyeva 1977)", and this may well have been the case in *Powichthys*. However, these tentative homologies remain open to debate. Furthermore, anteriorly to these processes, the dorsal edge of the commissural lamina is broken off, and a *processus ascendens* might have been present in this missing region. A large *processus ascendens* occurs in *Eusthenopteron* (Jarvik 1954, fig. 16C), *Glyptolepis* (Jarvik 1972, fig. 25) and, according to Jessen (1980), in the isolated palatoquadrate from the same locality as *Powichthys thorsteinssoni* (Jessen 1980, pl. 1:1). However, owing to the absence of a suprapterygoid process in *Powichthys*, it is reasonable to suppose that a *processus ascendens* is also absent or very reduced (and homologous to the "anterior" and "median" processes) in this genus, as in actinopterygians (Arratia & Schultze 1991). In this case the isolated palatoquadrate figured by Jessen (1980), which shows a strong *processus ascendens*, might belong to an undetermined "osteolepiform", and not to *Powichthys thorsteinssoni*. Nevertheless, if this isolated palatoquadrate actually belongs to *Powichthys thorsteinssoni*, we can assume that this large process was articulated with the skull roof, as in tetrapods, rather than with the braincase (Clack 1987; Ahlberg 1991).

The posterior margin of the *pars metapterygoidea* is very thick and filled with spongiose bone. The smooth dorsal surface of the commissural lamina shows numerous small depressions and small randomly arranged rounded bulges. In ventral view, the central part of the

ventral surface of the commissural lamina shows a raised platform for the tooth-bearing part of the entopterygoid (pf.pq, Fig. 8). It is impossible to determine whether this raised area, which is devoid of denticles, results from the fusion of the dorsal part of the entopterygoid to the palatoquadrate, or is a mere thickening of the commissural lamina proper. The anterior margin of this platform shows a strong overhang for the insertion of the posteromedial edge of the dermopalatine (Der, Fig. 10). More posteriorly and medially, the basal process of the palatoquadrate is very large and convex, and fits perfectly in the basipterygoid process of the sphenoid (bp.pr, Fig. 8).

Absence of a choana

Panchen & Smithson (1987) first defined the choana on the basis of the anatomy of fossil and living tetrapods, excluding functional, ontogenetic and phylogenetic inferences. They defined it as "an oval orifice bounded laterally by the premaxillary and/or maxillary and medially by the vomer and (dermo-)palatine". The presence of choanae was considered as a synapomorphy of the "Osteolepiformes" and tetrapods (Holmes 1985), although some authors have questioned their presence in the "Osteolepiformes" (Rosen *et al.* 1981), and the Rhizodontida are now assumed to possess choanae as well (Long & Ahlberg 1999). The presence of choanae in the Porolepiformes was previously claimed by numerous authors (e.g. Säve-Söderbergh 1933; Holmgren & Stensiö 1936; Romer 1937; Jarvik 1966, 1972, 1980; Bjerring 1989), but was subsequently challenged, and finally rejected (Rosen *et al.* 1981; Ahlberg 1991; Chang & Yu 1997; Forey 1998; Clément 2001).

Jessen (1975, 1980) interpreted *Powichthys thorsteinssoni* as a member of the Porolepiformes, and as a choanate. The *fenestra exochoanalis* is not observed in this species, because the maxilla, vomer and dermopalatine are lacking, but he considered the presence of a *fenestra ventrolateralis* (fe.vl, Fig. 13A), structurally close to that of the Porolepiformes, as evidence for a choana in *Powichthys*. The *fenestra ventrolateralis* of the Porolepiformes (and *Powichthys*) was supposed to have included a posterior *fenestra endonarina* and a "*fenestra endochoanalis*". In the Porolepiformes this "*fenestra endochoanalis*" is, however, closed ventrally by the ethmoidal process of the palatoquadrate, by the posterolateral expansion of the vomer, and by the anterolateral region of the dermopalatine (Clément 2001).

The holotype of *Powichthys spitsbergensis* displays the four dermal elements which bound the choana in the "Osteolepiformes" and tetrapods (and probably all Tetrapodomorpha, i.e. including rhizodonts); that is, the premaxilla, maxilla, vomer and dermopalatine. These dermal elements are adjacent to the left *fenestra ventrolateralis* of the specimen MNHN SVD 2156a, and are still in connection, although they have been very slightly displaced posteroventrally (Der, Max, Pmx, Vo, Figs. 7, 8). Their original position can be readily reconstructed, and they were clearly in close contact with each other, such that no orifice, even very small, could be present at their junction. It is thus certain that the choana was absent in *Powichthys spitsbergensis*, as in the Porolepiformes and, most probably, also *Powichthys thorsteinssoni*.

Post-orbital

The left post-orbital was found in connection with the skull (Po, Fig. 4A). It is thick, and longer than high, and its dorsal margin is neither straight, nor curved, as in the Porolepiformes, but sinuous, and thus fits perfectly in the irregular lateral margin of the anterior division of the skull. The anterior part of the post-orbital overlaps the skull roof (ov.Pa-eth, Fig. 11) and its posterior part is slightly overlapped by the first intertemporal, and to a large extent by the small dermal element situated in front of the overlap area for the first intertemporal. It is worth noticing that this condition is found in the Porolepiformes, and contrasts with that in other groups. In "Osteolepiformes" the anterior part of the post-orbital

is overlapped by the skull roof, but its posterior part overlaps the skull roof, whereas in *Youngolepis* the post-orbital is entirely overlapped by the skull roof (Chang 1982). The short posterodorsal margin of the post-orbital shows a broad overlap area (ov.Pre, Fig. 11). By comparison with *Powichthys thorsteinssoni* (Jessen 1980, fig. 3), one may assume that this area was probably overlapped by a pre-spiracular bone. The presence of this bone would be a synapomorphy shared by *Powichthys* and the Porolepiformes. The posteroventral margin of the post-orbital shows another overlap area (ov.Sq, Fig. 11) for a squamosal or a compound unit (due to the fusion of the pre-opercular, quadratojugal and squamosal), as in *Youngolepis* (Chang 1991) and some Middle Devonian "osteolepidids" (Jarvik 1948; Fan 1992; Chang & Zhu 1993). The anteroventral margin of the post-orbital forms an almost right angle with the posteroventral margin, and shows a very narrow overlap area for the jugal (ov.Ju, Fig. 11). The short orbital margin matches the orbital margin of the parieto-ethmoidal shield, confirming that the orbit was very small. The infra-orbital sensory-line canal passes dorsoventrally through the post-orbital. Its dorsal foramen is in a markedly posterior position, compared with that in the Porolepiformes and "Osteolepiformes". Its ventral foramen is situated at the posterior end of the overlap area for the jugal, also in an unusually posterior position.

Scales

Numerous scales, some articulated, have been found close to the skull in the same small block (Fig. 12A), and thus almost certainly belong to *Powichthys spitsbergensis*. The scales are cosmine covered and display an unusual shape, which differs from that in the "Porolepididae". They are rhombic in shape, with a very strong anteroventral process (av.pr, Fig. 12A–C), which is curved anterodorsally and digit-shaped. The scales figured by Jessen (1980, fig. 2B–E, pl. 4:6–9) lack this process, although at least one shows an anteroventral expansion (av.pr, Fig. 12D). A much less pronounced anteroventral process than that of *Powichthys spitsbergensis* occurs in the Chinese Middle Devonian osteolepids *Kenichthys campbelli* and *Thursius wudingensis* (av.pr, Fig. 12F, G), in scales referred by Tong-Dzuy & Janvier (1990) to *Youngolepis* sp. from the Lower Devonian of Vietnam (Fig. 12E), and in *Naxilepis*, a basal actinopterygian from the Late Silurian of China (Wang & Dong 1989).

The dorsal margin of the exposed area of the scale shows a very sigmoid shape. Contrary to *Powichthys thorsteinssoni* (Jessen 1980, fig. 2B–D) and the porolepidids *Porolepis* and *Heimenia* (Ørvig 1957, fig. 8D, F; 1969, fig. 5), there seems to be no evidence for ridges or tubercles anteriorly or dorsally to the cosmine layer. It can be noticed that the few, large, rounded tubercles

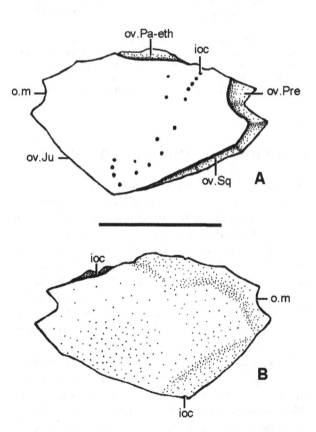

Fig. 11. Powichthys spitsbergensis sp. nov., Wood Bay Formation, Lower Devonian, northern Spitsbergen. Left post-orbital (holotype, MNHN SVD 2156b). A: External view. B: Internal view. Scale bar = 10 mm.

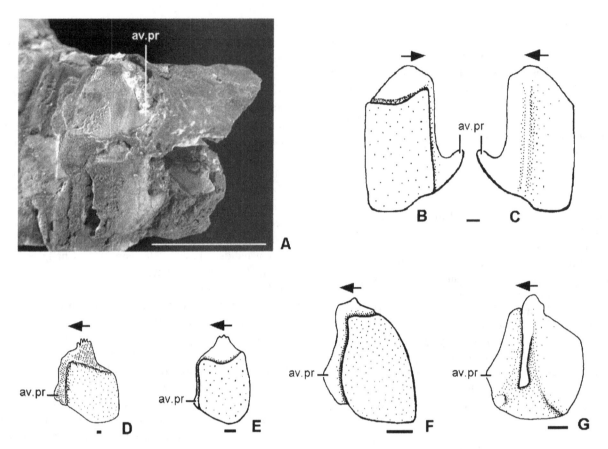

Fig. 12. Anteroventral process of scales. A–C: *Powichthys spitsbergensis* sp. nov., Wood Bay Formation, Lower Devonian, northern Spitsbergen (MNHN SVD 2156c, same block as the holotype). A: Part of squamation with scales in external view (scale bar = 10 mm). B, C: Scale in external and internal views (scale bar = 1 mm). D: *Powichthys thorsteinssoni* Jessen, 1975, scale in external view (after Jessen 1980). E: *Youngolepis* sp., scale in external view (after Tong-Dzuy & Janvier 1990, pl. 7:11). F: *Kenichthys campbelli* Chang & Zhu, 1993, scale in external view (after Chang & Zhu 1993, fig. 14A, E). G: *Thursius wudingensis* Fan, 1992, scale in internal view (after Fan 1992, fig. 7B). Scale bars = 1 mm. The black arrow points forwards.

found on the scales of *Powichthys thorsteinssoni* (personal observation of the material housed in the Georg-August University of Göttingen, Germany) are different from the spoon-shaped dentine tubercles seen in various Porolepiformes, e.g. *Heimenia, Laccognathus, Glyptolepis, Holoptychius,* and *Quebecius* (Ørvig 1957; Jarvik 1980; Cloutier & Schultze 1996), in certain Dipnoi such as *Iowadipterus* or *Tarachomylax* (Schultze 1992; Barwick *et al.* 1997), and in the onychodontids *Onychodus, Strunius,* and *Grossius* (Ørvig 1957; Jessen 1967; Schultze 1973; also an undetermined onychodont from Morocco; Aquesbi 1988).

The internal surface of *Powichthys spitsbergensis* scales is smooth and slightly concave (Fig. 12C). A straight thin crest is close to the anterior margin of the scale. The very anterior position of this crest is unusual, and quite different from its position in the scales of the Porolepiformes and "Osteolepiformes". No trace of digitations of the dorsal overlap area of the scales has been found in the Spitsbergen material, contrary to *Powichthys thorsteinssoni* (Jessen 1980, fig. 2B, C).

Conclusions and summary

The morphology of this new *Powichthys* species from the Lower Devonian of Spitsbergen allows us to reinterpret some conjectural features previously assumed by Jessen (1975, 1980) for *Powichthys thorsteinssoni*. Thus, the junction between the supra-orbital and infra-orbital sensory-line canals is in fact lacking, the isolated palatoquadrate described from the Lower Devonian of Arctic Canada and referred to *Powichthys thorsteinssoni* is not porolepiform-like but osteolepiform-like, and the presumed choana is clearly absent in the Powichthyidae, as was also recently shown in the Porolepiformes (Clément 2001). Some features observed in *Powichthys spitsbergensis* confirm that this genus is not a porolepiform, but displays a mixture of characters that were regarded as unique to either osteolepiforms, porolepiforms, or dipnoans, along-side general sarcopterygian characters (e.g. the presence of a *processus dermintermedius*, a polyplocodont tooth structure, the presence of a descending process of the basisphenoid, premaxillae independent from the

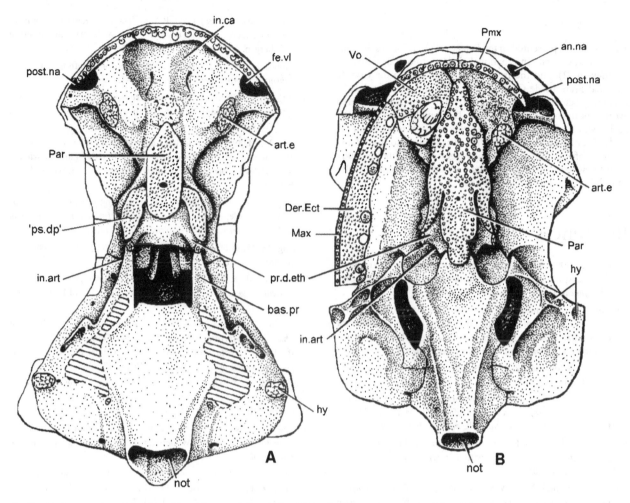

Fig. 13. Braincase and overlying dermal bones in ventral view. A: *Powichthys thorsteinssoni* Jessen, 1975 (after Janvier 1996, modified from Jessen 1980). B: *Youngolepis praecursor* Chang & Yu, 1981 (after Janvier 1996, modified from Chang 1982). Not to scale.

parieto-ethmoidal shield and not traversed by the infra-orbital sensory-line canal, the floor of the internasal cavities displaying a network of interconnected grooves, an osteolepiform-like palatoquadrate, and the absence of a junction between the supra-orbital and infra-orbital sensory-line canals).

This assemblage of characters makes the phylogenetic position of *Powichthys* still highly debated. Ahlberg (1991), Cloutier & Ahlberg (1996), and Zhu *et al.* (2001) interpreted *Powichthys* as the sister taxon of all other Dipnoiformes (*sensu* Cloutier 1990); Forey (1998) considered it as the sister taxon of all other Dipnomorpha, whereas Zhu & Schultze (1997, 2001) considered it as the sister taxon of all other Sarcopterygii except *Youngolepis*, *Diabolepis* and the Dipnoi.

The excellent preservation of the new material described here also provides details of some new morphological features. Notable among these are the strong, dorsally concave, internal process of the lacrimal (also present in *Porolepis*), extending from the posterior part of the posterior nostril to the anterior margin of the orbit,

and strikingly reminiscent of the nasolacrimal duct of tetrapods; the large palatal dental plates covering the buccal roof and probably correlated with a crushing palatal bite; the post-orbital bone, previously unknown, which shows an unusual path of the infra-orbital sensory-line canal compared with other dipnomorphs; and the scales with their very strong anteroventral processes, unknown in other osteichthyans.

The new species *Powichthys spitsbergensis* is also of palaebiogeographical interest. Five of the seven oldest known sarcopterygians (Late Silurian and Lochkovian–Pragian) come from south China and north Vietnam: *Achoania* Zhu *et al.*, 2001, *Psarolepis* Yu, 1998, *Diabolepis* Chang & Yu, 1984 (Chang 1995), *Youngolepis* Chang & Yu, 1981 (Chang 1982, 1991; Tong-Dzuy & Janvier 1990), and *Langdenia* Janvier & Ta Hoa, 1999. However, the occurrence of *Powichthys* in Arctic Canada and Spitsbergen, and the widespread occurrence of the "Porolepididae" in Europe, both challenge the hypothesis that the South China Block may have been the centre of origin for the Sarcopterygii (Zhu *et al.* 2001).

Acknowledgements

This paper was presented to the Palaeozoic Vertebrates Symposium of the First International Palaeontological Congress (Sydney, 2002), and the first author is grateful to the Comité National Français de Géologie and the Centre National de la Recherche Scientifique, Paris, for providing funding for travel to Sydney. Both authors are grateful to Dr Daniel Goujet (MNHN, Paris) who passed this new *Powichthys* material on to them for study. Thanks are also due to Dr Per-Erik Ahlberg for useful discussion, information and helpful suggestions during the study of this new material. Photographs were taken by Denis Serrette and Philippe Loubry, MNHN, Paris.

References

Ahlberg, P.E. 1991: A re-examination of sarcopterygian interrelationships, with special reference to the Porolepiformes. *Zoological Journal of the Linnean Society, London 103*, 241–287.

Ahlberg, P.E. & Johanson, Z. 1997: Second tristichopterid (Sarcopterygii, Osteolepiformes) from the Upper Devonian of Canowindra, New South Wales, Australia, and phylogeny of the tristichopteridae. *Journal of Vertebrate Paleontology 17(4)*, 653–673.

Ahlberg, P.E. & Johanson, Z. 1998: Osteolepiformes and the ancestry of tetrapods. *Nature 395*, 792–794.

Aquesbi, N. 1988: Etude d'un Onychodontiforme (Osteichthyes, Sarcopterygii) du Dévonien moyen (Eifelien) du Maroc. *Bulletin du Muséum National d'Histoire Naturelle de Paris 10*, 181–196.

Arratia, G. & Schultze, H.-P. 1991: Palatoquadrate and its ossifications: development and homology within osteichthyans. *Journal of Morphology 208*, 1–81.

Barwick, R.E., Campbell, K.S.W. & Mark-Kurik, E. 1997: *Tarachomylax:* a new Early Devonian dipnoan from Severnaya Zemlya, and its place in the evolution of Dipnoi. *Geobios 30(1)*, 45–73.

Bjerring, H.C. 1975: Contribution à la connaissance de la neuro-épiphyse chez les urodèles et leurs ancêtres porolépiformes avec quelques remarques sur la signification évolutive des muscles striés parfois présents dans la région neuro-épiphysaire des mammifères. *In* Lehman, J.P. (ed.): *Problèmes actuels de Paléontologie. Evolution des Vertébrés*, Vol. 1, 231–256. *Colloques Internationaux du Centre National de la Recherche Scientifique 218*. Editions du CNRS, Paris.

Bjerring, H.C. 1989: Apertures of craniate olfactory organs. *Acta Zoologica 70(2)*, 71–85.

Bjerring, H.C. 1994: The evolutionary origin and homologues of the supracochlear lamina: a contribution to our knowledge of mammalian ancestry. *Acta Zoologica 75(4)*, 359–369.

Bjerring, H.C. 1995: The question of a homology between the reptilian *processus basipterygoideus* and the mammalian *processus alaris*. *Palaeontographica 235*, 79–96.

Blieck, A. 1984: Les Hétérostracés Ptéraspidiformes, Agnathes du Silurien-Dévonien du Continent nord-atlantique et des blocs avoisinants: révision systématique, phylogénie, biostratigraphie, biogéographie. *Cahiers de Paléontologie (Vertébrés)*. Editions du CNRS, Paris.

Blieck, A. & Cloutier, R. 2000: Biostratigraphical correlations of Early Devonian vertebrate assemblages of the Old Red Sandstone Continent. *In* Blieck, A. & Turner, S. (eds): *Palaeozoic Vertebrate Biochronology and Global Marine/Non-marine Correlation. Final Report of IGCP 328 (1991–1996)*, 223–270. Courier Forschungsinstitut Senckenberg, Frankfurt.

Blieck, A., Goujet, D. & Janvier, P. 1987: The vertebrate stratigraphy of the Lower Devonian (Red Bay Group and Wood Bay Formation) of Spitsbergen. *Modern Geology 11*, 197–217.

Campbell, K.S.W. & Barwick, R.E. 1987: Paleozoic lungfishes – a review. *Journal of Morphology Supplement 1*, 93–131.

Chang, M.-M. 1982: The braincase of *Youngolepis*, a Lower Devonian crossopterygian from Yunnan, south-western China. PhD thesis. GOTAB, University of Stockholm.

Chang, M.-M. 1991: Head skeleton and shoulder girdle of *Youngolepis*. *In* Chang, M.-M., Liu, Y.-H. & Zhang, G.-R. (eds): *Early Vertebrates and Related Problems of Evolutionary Biology*, 355–378. Science Press, Beijing.

Chang, M.-M. 1995: *Diabolepis* and its bearing on the relationships between porolepiforms and dipnoans. *Bulletin du Muséum National d'Histoire Naturelle, Paris, 4e série, Section C 17*, 235–268.

Chang, M.-M. & Smith, M.M. 1992: Is *Youngolepis* a porolepiform? *Journal of Vertebrate Paleontology 12*, 294–312.

Chang, M.-M. & Yu, X. 1981: A new crossopterygian, *Youngolepis praecursor*, gen. et sp. nov., from Lower Devonian of East Yunnan, China. *Scientia Sinica 24*, 89–97.

Chang, M.-M. & Yu, X. 1984: Structure and phylogenetic significance of *Diabolichthys speratus* gen. et sp. nov., a new Dipnoan-like form from the Lower Devonian of Eastern Yunnan, China. *Proceedings of the Linnean Society of New South Wales 107(3)*, 171–184.

Chang, M.-M. & Yu, X. 1997: Reexamination of the relationship of Middle Devonian Osteolepids – fossil characters and their interpretations. *American Museum Novitates 3189*, 1–20.

Chang, M.-M. & Zhu, M. 1993: A new Middle Devonian osteolepid from Qujing, Yunnan. *Memoirs of the Association of Australasian Palaeontologists 15*, 183–198.

Clack, J.A. 1987: *Pholiderpeton scutigerum* Huxley, an amphibian from the Yorkshire coal measures. *Philosophical Transactions of the Royal Society of London (B) 318(1188)*, 1–107.

Clément, G. 2001: Evidence for lack of choanae in the Porolepiformes. *Journal of Vertebrate Paleontology 21(4)*, 795–802.

Cloutier, R. 1990: Phylogenetic interrelationships of the Actinistians (Osteichthyes: Sarcopterygii): patterns, trends, and rates of evolution. PhD thesis. University of Kansas, Lawrence.

Cloutier, R. 1996: Dipnoi (Akinetia: Sarcopterygii). *In* Schultze, H.-P. & Cloutier, R. (eds): *Devonian Fishes and Plants from Miguasha, Quebec, Canada*, 198–226. Verlag Dr Friedrich Pfeil, München.

Cloutier, R. 1997: Morphologie et variations du toit crânien du dipneuste *Scaumenacia curta* (Whiteaves) (Sarcopterygii), du Dévonien supérieur du Québec. *Geodiversitas 19(1)*, 61–105.

Cloutier, R. & Ahlberg, P.E. 1996: Morphology, characters, and the interrelationships of the basal sarcopterygians. *In*: Stiassny, M.L.J., Parenti, L. & Johnson, G.D. (eds.): *Interrelationships of Fishes II*, 445–479. Academic Press, New York.

Cloutier, R. & Schultze, H.-P. 1996: Porolepiform fishes (Sarcopterygii). *In* Schultze, H.-P. & Cloutier, R. (eds): *Devonian Fishes and Plants from Miguasha, Quebec, Canada*, 248–270. Verlag Dr Friedrich Pfeil, München.

Dineley, D.L. & Loeffler, E.J. 1993: Biostratigraphy of the Silurian and Devonian gnathostomes of the Euramerican Province. *In* Long, J.A. (ed.): *Palaeozoic Vertebrate Biostratigraphy and Biogeography*, 104–138. Belhaven Press, London.

Elliott, D.K. 1984: A new subfamily of the Pteraspididae (Agnatha: Heterostraci) from the Upper Silurian and Lower Devonian of Arctic Canada. *Palaeontology 27*, 169–197.

Elliott, D.K. & Dineley, D.L. 1983: New species of *Protopteraspis* (Agnatha, Heterostraci) from the (?)Upper Silurian to Lower Devonian of Northwest Territories, Canada. *Journal of Paleontology 57(3)*, 474–494.

Elliott, D.K. & Dineley, D.L. 1991: Additional information on *Alainaspis* and *Boothiaspis*, cyathaspidids (Agnatha: Heterostraci) from the Upper Silurian of Northwest Territories, Canada. *Journal of Paleontology 65(2)*, 308–313.

Elliott, D.K., Loeffler, E.J. & Liu, Y. 1998: New species of the cyathaspidid *Poraspis* (Agnatha: Heterostraci) from the Late Silurian and Early Devonian of Northwest Territories, Canada. *Journal of Paleontology 72(2)*, 360–370.

Fan, J. 1992: A new species of *Thursius* from Wuding, Yunnan. *Vertebrata PalAsiatica 30*, 195–209 (in Chinese).

Forey, P.L. 1998: *History of the Coelacanth Fishes*. Chapman & Hall, London.

Gardiner, B.G. 1984: The relationships of the palaeoniscid fishes, a review based on new specimens of *Mimia* and *Moythomasia* from the Upper Devonian of Western Australia. *Bulletin of the British Museum (Natural History), Geology 37(4)*, 173–428.

Goujet, D. 1984: Les poissons placodermes du Spitsberg: Arthrodires Dolichothoraci de la Formation de Wood Bay (Dévonien inférieur). *Cahiers de Paléontologie*. Editions du CNRS, Paris.

Holmes, E.B. 1985: Are lungfishes the sister group of tetrapods? *Biological Journal of the Linnean Society 25*, 379–397.

Holmgren, N. & Stensiö, E.A. 1936: Kranium und visceralskelett der Akranier, Cyclostomen unf Fische. *In* Bolk, L., Göppert, E., Kallius, E. & Lubosch, W. (eds): *Handbuch der Vergleichenden Anatomie der Wirbertiere 4*, 233–500. Berlin.

Janvier, P. 1996: *Early Vertebrates*. Oxford Science Publication, Clarendon Press.

Janvier, P. & Blieck, A. 1993: The Silurian–Devonian agnathan biostratigraphy of the Old Red Continent. *In* Long, J.A. (ed.): *Palaeozoic Vertebrate Biostratigraphy and Biogeography*, 67–86. Belhaven Press, London.

Janvier, P. & Ta Hoa, P. 1999: Les vertébrés (Placodermi, Galeaspida) du Dévonien inférieur de la coupe de Lung Cỏ-Mia Lé, province de Hà Giang, Viêt Nam, avec des données complémentaires sur les gisements à vertébrés du Dévonien du Bac Bo oriental. *Geodiversitas 21(1)*, 33–67.

Jarvik, E. 1942: On the structure of the snout of crossopterygians and lower gnathostomes in general. *Zoologiska Bidrag fran Uppsala 21*, 235–675.

Jarvik, E. 1948: On the morphology and taxonomy of the Middle Devonian osteolepid fishes of Scotland. *Kungliga Svenska Vetenskapsakademiens Handlingar 25(1)*, 1–301.

Jarvik, E. 1954: On the visceral skeleton in *Eusthenopteron* with a discussion of the parasphenoid and palatoquadrate in fishes. *Kungliga Svenska Vetenskapsakademiens Handlingar 5*, 1–104.

Jarvik, E. 1966: Remarks on the structure of the snout in *Megalichthys* and certain other rhipidistid crossopterygians. *Arkiv för Zoologi 19*, 41–98.

Jarvik, E. 1972: Middle and Upper Devonian Porolepiformes from East Greenland with special reference to *Glyptolepis groenlandica* n. sp., and a discussion on the structure of the head in the Porolepiformes. *Meddelelser om Grønland 187(2)*, 1–307.

Jarvik, E. 1980: *Basic Structure and Evolution of Vertebrates*, Vol. 1. Academic Press, London.

Jessen, H.L. 1967: The position of the Struniiformes *Strunius* and *Onychodus* among the crossopterygians. *In* Lehman, J.-P. (ed.): *Problèmes actuels de Paléontologie. Evolution des Vertébrés*, 173–178.

Colloques Internationaux du Centre National de la Recherche Scientifique 163. Editions du CNRS, Paris.

Jessen, H.L. 1975: A new choanate fish, *Powichthys thorsteinssoni* n.g., n.sp., from the Early Lower Devonian of the Canadian Arctic Archipelago. *In* Lehman, J.P. (ed.): *Problèmes actuels de Paléontologie. Evolution des Vertébrés*, 213–222. *Colloques Internationaux du Centre National de la Recherche Scientifique 218*. Editions du CNRS, Paris.

Jessen, H.L. 1980: Lower Devonian Porolepiformes from the Canadian Arctic with special reference to *Powichthys thorsteinssoni* Jessen. *Palaeontographica A 167*, 180–214.

Johanson, Z. & Ahlberg, P.E. 1997: A new tristichopterid (Osteolepiformes: Sarcopterygii) from the Mandagery Sandstone (Late Devonian, Famennian) near Canowindra, NSW, Australia. *Transactions of the Royal Society of Edinburgh: Earth Sciences 88*, 39–68.

Lebedev, O.A. 1995: Morphology of a new osteolepidid fish from Russia. *Bulletin du Muséum National d'Histoire Naturelle, Paris, 4e série, Section C 17*, 287–341.

Long, J.A. & Ahlberg, P.E. 1999. New observations on the snouts of rhizodont fishes (Palaeozoic Sarcopterygii). *Records of the Western Australian Museum 57*, 169–173.

Märss, T., Caldwell, M., Gagnier, P.-Y., Goujet, D., Männik, P., Martma, T. & Wilson, M. 1998: Distribution of Silurian and Lower Devonian vertebrate microremains and conodonts in the Baillie-Hamilton and Cornwallis Island sections, Arctic Canada. *Proceedings of the Estonian Academy of Sciences, Geology 47(2)*, 51–76.

Miles, R.S. 1977: Dipnoan (lungfish) skulls and the relationships of the group: a study based on new species from the Devonian of Australia. *Zoological Journal of the Linnean Society 61*, 1–328.

Ørvig, T. 1957: Remarks on the vertebrate fauna of the Lower Upper Devonian of Escuminac Bay, P. Q., Canada, with special reference to the Porolepiform Crossopterygians. *Arkiv för Zoologi 10(6)*, 367–426.

Ørvig, T. 1969: Vertebrates from the Wood Bay Group and the position of the Emsian–Eifelian boundary in the Devonian of Vestspitsbergen. *Lethaia 2*, 273–328.

Panchen, A.L. & Smithson, T.R. 1987: Character diagnosis, fossils and the origin of tetrapods. *Biological Reviews 62*, 341–438.

Romer, A.S. 1937: The braincase of the carboniferous crossopterygian *Megalichthys nitidus*. *Bulletin of the Museum of Comparative Zoology, Harvard 82*, 1–73.

Rosen, D.E., Forey, P.L., Gardiner, B.G. & Patterson, C. 1981: Lungfishes, tetrapods, paleontology, and plesiomorphy. *Bulletin of the American Museum of Natural History 167(4)*, 159–276.

Säve-Söderbergh, G. 1933: The dermal bones of the head and the lateral-line system in *Osteolepis macrolepidotus* Agassiz with remarks on the terminology of the lateral-line system and on the dermal bones of certain other crossopterygians. *Nova Acta of the Royal Society of Sciences of Uppsala 9(2)*, 1–130.

Schultze, H.-P. 1973: Crossopterygier mit heterozerker schwanzflosse aus dem Oberdevon Kanadas, nebst einer beschreibung von Onychodontida-resten aus dem Mitteldevon Spaniens und aus dem Karbon der USA. *Palaeontographica 143*, 188–208.

Schultze, H.-P. 1992: A new long-headed Dipnoan (Osteichthyes) from the Middle Devonian of Iowa, USA. *Journal of Vertebrate Paleontology 12(1)*, 42–58.

Thorsteinsson, R. 1967: Preliminary note on Silurian and Devonian ostracoderms from Cornwallis and Somerset Islands, Canadian Arctic Archipelago. *In* Lehman, J.-P. (ed.): *Problèmes actuels de Paléontologie. Evolution des Vertébrés*, 45–47. *Colloques Internationaux du Centre National de la Recherche Scientifique 163*. Editions du CNRS, Paris.

Tong-Dzuy, T. & Janvier, P. 1990: Les vertébrés du Dévonien inférieur du Bac Bo oriental (provinces de Bac thaï et Lang Son, Viêt Nam). *Bulletin du Muséum National d'Histoire Naturelle, Paris 12(2)*, 143–223.

Vieth, J. 1980: Thelodontier-, Acanthodier-, und Elasmobranchier-Schuppen aus dem Unter-Devon der kanadischen Arktis (Agnatha, Pisces). *Göttingen Arbeiten in Geologie und Paläontologie 23*, 1–69.

Vorobyeva, E.I. 1977: Morfologiya i osobennosti evolyutsii kisteperykh ryb. [Morphology and nature of evolution of crossopterygian fishes.] *Trudy Paleontologicheskogo Instituta 163*, 1–239 (in Russian).

Wang, N.-Z. & Dong, Z.-H. 1989: Discovery of Late Silurian microfossils of Agnatha and fishes from Yunnan, China. *Acta Palaeontologica Sinica 28*, 192–206 (in Chinese).

Young, G.C. 1993: Vertebrate faunal provinces in the Middle Palaeozoic. *In* Long, J.A. (ed.): *Palaeozoic Vertebrate Biostratigraphy and Biogeography*, 293–323. Belhaven Press, London.

Yu, X. 1998: A new porolepiform-like fish, *Psarolepis romeri*, gen. et sp. nov. (Sarcopterygii, Osteichthyes) from the Lower Devonian of Yunnan, China. *Journal of Vertebrate Paleontology 18(2)*, 261–274.

Zhu, M. & Schultze, H.-P. 1997: The oldest sarcopterygian fish. *Lethaia 30*, 293–304.

Zhu, M. & Schultze, H.-P. 2001: Interrelationships of basal osteichthyans. *In* Ahlberg, P.E. (ed.): *Major Events in Early Vertebrate Evolution*, 289–314. Taylor and Francis, London.

Zhu, M., Yu, X. & Ahlberg, P.E. 2001: A primitive sarcopterygian fish with an eyestalk. *Nature 410*, 81–84.

A terrestrial vertebrate assemblage from the Late Palaeozoic of central Germany, and its bearing on Lower Permian palaeoenvironments

STUART S. SUMIDA, DAVID S. BERMAN, DAVID A. EBERTH & AMY C. HENRICI

Sumida, S.S., Berman, D.S., Eberth, D.A. & Henrici, A.C. 2004 06 01. A terrestrial vertebrate assemblage from the Late Palaeozoic of central Germany, and its bearing on Lower Permian palaeoenvironments. *Fossils and Strata*, No. 50, pp. 113–123. USA. ISSN 0300-9491.

The Bromacker locality of the Lower Permian Tambach Formation, lowermost unit of the Upper Rotliegend of central Germany, is unique among localities of comparable age. It preserves an exclusively terrestrial assemblage. Its unique nature allows an assessment of the degree of terrestriality of other Lower Permian assemblages, and the Bromacker assemblage contributes to efforts to develop a terrestrial vertebrate basis for recognising the base of the Permian globally. Previous comparisons between the Bromacker and other Lower Permian localities have been a mixture of locality-specific and more broadly regional comparisons. A more accurate assessment of the palaeoecological and biostratigraphic significance of the Bromacker, from a comparison with specific localities for which confident palaeoenvironmental interpretations are available, demonstrates its strictly terrestrial nature, and shows the Bromacker assemblage to provide a standard by which terrestrial taxa may be distinguished from semi-terrestrial and aquatic taxa at other localities. The seymouriamorph amphibian *Seymouria* proves to be a useful taxon for terrestrial biostratigraphic correlation in the Lower Permian.

Key words: Early Permian; terrestrial biostratigraphy; palaeoenvironments; *Seymouria*; *Dimetrodon*; Diadectomorpha.

Stuart S. Sumida [ssumida@csusb.edu], Department of Biology, California State University San Bernardino, 5500 University Parkway, San Bernardino, CA 92506, USA

David S. Berman [bermand@carnegiemuseums.org], Section of Vertebrate Paleontology, Carnegie Museum of Natural History, 4400 Forbes Avenue, Pittsburgh, PA 15213, USA

David A. Eberth [David.Eberth@gov.ab.ca.], Royal Tyrrell Museum of Palaeontology, P.O. Box 7500, Drumheller, Alberta TOJ 0Y0, Canada

Amy C. Henrici [henricia@carnegiemuseums.org], Section of Vertebrate Paleontology, Carnegie Museum of Natural History, 4400 Forbes Avenue, Pittsburgh, PA 15213, USA

Introduction

The Lower Permian Bromacker locality of the Tambach Formation in central Germany (Fig. 1) has been known as an important locality for Early Permian age footprints for over a century (Pabst 1896). More recently, Martens (1988) expanded the palaeobiological investigation at the locality beyond ichnofossils, and over the last decade a series of studies has documented a moderately sized assemblage of extremely well-preserved vertebrate fossils (Berman & Martens 1993; Berman *et al.* 1998, 2000). The faunal list from this vertebrate assemblage as currently known is given in Table 1.

Most recently, environmental, sedimentological, and palaeoecological analyses (Eberth *et al.* 2000) have demonstrated clearly that the locality differed significantly from some well-known non-marine vertebrate localities of the Lower Permian of North America. Most North American localities of Early Permian age preserve mixed assemblages of aquatic, semi-aquatic, and terrestrial taxa in varying proportions. These assemblages are generally preserved in red-bed deposits that reflect environments of deposition with significant, but not exclusively, aquatic and semi-aquatic components (Table 2). In contrast, most European vertebrate localities of the Early Permian are limnetic grey shales reflective of a more exclusively

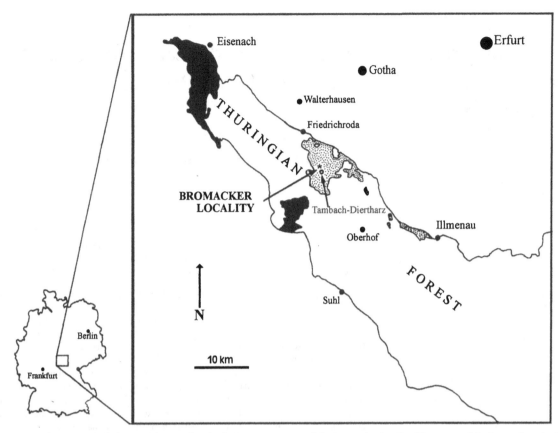

Fig. 1. A map of Germany with the inset showing the Thuringian Forest area and the Bromacker locality. The extent of the Lower Permian Tambach Formation is indicated by stippling. Solid black indicates the extent of other Lower Permian strata (primarily the Eisenach Formation).

aquatic environment. The paleoenvironment of the Bromacker locality stands in strong contrast to other European localities of comparable age as a red-bed deposit, and recent studies (Berman & Martens 1993; Sumida *et al.* 1996, 1998b; Berman *et al.* 1998, 2000a, 2001) have demonstrated that certain components of its preserved vertebrate assemblage were more similar to North American red-bed assemblages than to those limnetic assemblages commonly found in the Lower Permian of most of Europe. Thus, the Bromacker locality assemblage provides an opportunity to assess the models of Early Permian red-bed paleoenvironments that had, until very recently, been developed exclusively on the basis of data derived from North American red-bed localities. In those interpretations, terrestrial palaeoeco-systems are thought to have occupied extensive lowland coastal and alluvial plain settings (e.g. Olson 1977; Sander 1987, 1989; Eberth & Miall 1991). Although upland settings and fossil assemblages are important to a complete understanding of the evolution and palaeocology of Late Palaeozoic tetrapods, they are not as well documented (Vaughn 1966, 1969, 1972; Eberth & Berman 1993; Eberth *et al.* 2000).

If the Bromacker locality is to be a useful measure of comparison to vertebrate assemblages of the Early

Table I. Vertebrate assemblage known from the Lower Permian Bromacker locality, Thuringian Forest, central Germany.

"Amphibia"
    Temnospondyli
        Dissorophoidea
            New taxon
        Trematopidae
            Tambachia trogallas (Sumida *et al.* 1998b)
            New taxon
    Seymouriamorpha
        Seymouria sanjuanensis (Berman & Martens 1993)

Amniota(?)
    Diadectomorpha
        Diadectidae
            Diadectes absitus (Berman *et al.* 1998)
            New Taxon (Kissel *et al.* 2002)

Amniota
    Synapsida
        Caseidae
            New Taxon (Sumida *et al.* 2001)
        Varanopseidae
            New Taxon? (Sumida *et al.* 2001)
        Sphenacodontidae
            Dimetrodon teutonis (Berman *et al.* 2001)
    Reptilia
        Protorothyrididae
            Thuringothyris mahlendorffae (Boy & Martens 1991)
        Bolosauridae
            Eudibamus cursoris (Berman *et al.* 2000b)

Table II. Palaeoenvironmental interpretations for Late Pennsylvanian and Early Permian localities in North America and central Germany.

Locality	Environmental interpretation	Reference
Late Pennsylvanian		
Badger Creek, Colorado	Pond or oxbow lake.	Vaughn 1969
El Cobre Canyon, New Mexico	Anastomosing streams, overbank deposits of fluvial origin.	Eberth & Miall 1991
Halgaito Sandstone, Utah	Anastomosing streams in arid coastal lowlands	Baars 1962
Early Permian		
Organ Rock Shale, Utah	Anastomosing streams in arid coastal lowlands or floodplain.	Baars 1962
Arroyo del Agua, New Mexico	Anastomosing fluvial channels and crevasse splay deposits.	Eberth & Miall 1991
Placerville, Colorado		
Rattlesnake Canyon, Texas	Pond or oxbow lake.	Sander 1989
Archer City, Texas	Pond or oxbow lake.	Sander 1989
Geraldine Bonebed, Texas	"*Wasserleichen*", subsiding flood basinal environment.	Sander 1987
Waurika, Oklahoma	Large pond or lake deposit.	Olson 1977
Prince Edward Island	Aggrading stream channels in deltaic deposits.	Langston 1963
Fort Sill, Oklahoma	Predominantly terrestrial karst deposit.	Reisz & Sutherland 2001
Bromacker, Germany	Upland, exclusively terrestrial, internally drained palaeograben.	Eberth *et al.* 2000

Permian of North America, then its faunal (biological) and geological characteristics must be clearly defined. In both cases, the locality appears to reflect one of the most strictly terrestrial red-bed localities known from the Lower Permian. This interpretation is supported by both sedimentological and palaeontological analyses of the locality. Eberth *et al.* (2000) showed that the Tambach Basin was, at the time of deposition of the Bromacker locality, a small and internally drained area in an upland setting near the centre of a small, isolated, internally drained basin, far removed and up-dip from regional drainage systems associated with coastal or alluvial plains. This is in contrast to almost all North American localities that have been analysed from the Late Palaeozoic. Depositional events at the Bromacker were dominated by seasonal to subseasonal cycles of flooding. Flood events took place in an ephemeral, alluvial-to-lacustrine setting that experienced annual precipitation similar to that of a wet-and-dry (or wetter) tropical climate, which was hot throughout the year. Beyond periodic flood events, it appears that high-energy or continuously active stream channels did not persist at the Bromacker due to its isolation from regional drainage systems. The presence of virtually undisturbed complete vertebrate skeletons, and uniform mud and clay drapes deposited over shallow submerged palaeodunes, are evidence of the calm to still-standing nature of the transiently present water in the internally drained basin (Eberth *et al.* 2000).

In terms of the evidence provided by those vertebrate fossils, the Bromacker locality shares a number of taxa in common with well-documented localities known from the Late Palaeozoic of North America. However, in concert with its unusual preservation of a rarely analysed palaeoecological situation, its fossil vertebrate assemblage

is also distinct from those North American localities in a number of ways (Berman *et al.* 2000; Eberth *et al.* 2000). First, a majority of the taxa known from the Bromacker are extremely well preserved (Fig. 2), suggesting little or no transport, and that death and burial were probably coeval events. Second, the assemblage is dominated by terrestrial high-fibre herbivores including *Diadectes*, plus a new taxon that is closely related to *Diadectes*, a new caseid pelycosaur, and the bolosaurid reptile *Eudibamus*. Third, in terms of numbers of specimens, pelycosaurian-grade synapsids are relatively more rare than diadectids, and finally there is a complete lack of aquatic or semi-aquatic vertebrates. The last feature is significant in its relationship to the sedimentological summary presented above, in that the interpretation of the Tambach Basin as small and internally drained suggests that there was probably little opportunity for aquatic or semi-aquatic vertebrates to colonise the basin. Taken together, the biological and sedimentological interpretations suggest that the Bromacker locality provides the earliest evidence of a truly upland, terrestrial ecosystem that had at its base a dominant stock of high-fibre herbivores (Berman *et al.* 2000; Eberth *et al.* 2000). These studies suggested further that the Bromacker provides evidence that experiments with herbivores as the dominant or significant basal component of vertebrate food webs had begun by the Early Permian, in ecological settings devoid of aquatic or semi-aquatic vertebrates.

The confident interpretation of the Bromacker locality as an upland, terrestrial ecosystem allows it to bear on both biostratigraphic and palaeoecological comparisons with other Lower Permian localities worldwide. Significantly, the well-known Lower Permian red-bed localities of North America have all been interpreted as having at

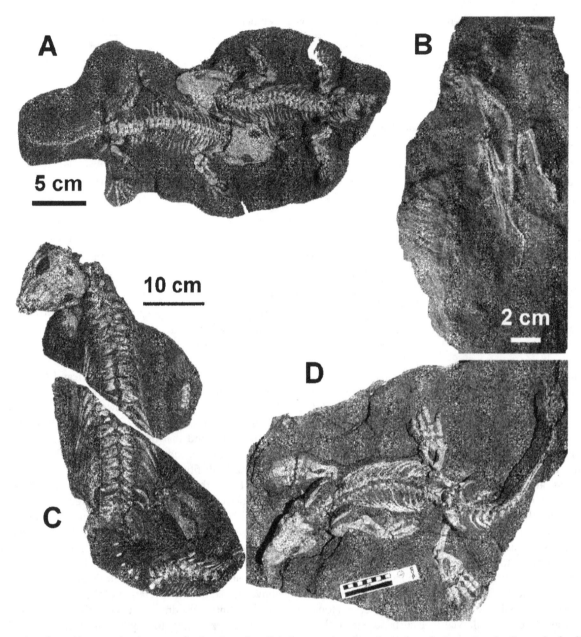

Fig. 2. Selected taxa demonstrating the taxonomic diversity and quality of preservation of specimens from the Lower Permian Bromacker locality of central Germany. A: Two complete skeletons of the seymouriamorph tetrapod *Seymouria sanjuanensis.* B: The facultative bipedal bolosaurid, *Eudibamus cursoris.* C: The diadectomorph tetrapod *Diadectes absitus.* D: *Orobates pabsti*, a new diadectomorph tetrapod closely related to *Diadectes* (bar scale in D = 10 cm).

least some aquatic or semi-aquatic components (Olson 1977; Sander 1987, 1989; Sumida *et al.* 1996; Eberth *et al.* 2000). A comparison of the Bromacker and its vertebrate assemblage with those of other Late Palaeozoic localities could allow the identification of terrestrial components of the assemblages, essentially allowing them to be "dissected out" of the larger assemblage. This could in turn facilitate their distinction from taxa that are more likely aquatic or semi-aquatic. Finally, recent attempts to develop global subdivisions of the Permian that are independent of the marine record (Lucas 2002) further

emphasise the importance of the Bromacker, as it is the only Lower Permian locality that contains taxa that can be confidently described as strictly terrestrial, and thus clearly without any marine influence.

Comparative geological and geographical contexts

Previous comparisons of the Bromacker locality with Lower Permian assemblages in North America have involved both specific localities and broader geographical regions (Sumida *et al.* 1999, 2000). A more rigorous method of comparison would be on a locality-by-locality basis, avoiding the potential lumping of multiple environments of deposition into a single category. Here, assemblages from individual localities of Early Permian age are compared with those of the Bromacker whenever possible. When not possible, as restricted a comparison as possible is attempted. In an attempt to refine the biostratigraphic utility of the taxa found at the Bromacker locality, well-documented assemblages of Late Pennsylvanian age are also included in the comparison. Localities in the Valley of the Gods and John's Canyon in southeastern Utah are known from the Halgaito Shale of the Cutler Formation in southeastern Utah. Although previously interpreted as Early Permian (Baker 1936; Vaughn 1962), more recent interpretations of the invertebrate assemblages of stratigraphic equivalents in this area (Baars 1991, 1995; Chernykh *et al.* 1997; Davydov *et al.* 1995) suggest a Late Pennsylvanian assignment. Analyses of the vertebrate assemblages of the above localities have followed this latter interpretation (Sumida *et al.* 1998a, c, d). In this analysis, the assemblages for John's Canyon and the Valley of the Gods are combined, as their individual assemblage components are similar, and they have been interpreted as similar in age (Sumida *et al.* 1998a, c, d) and depositional environment (Frede *et al.* 1993). Badger Creek locality near Howard, Colorado (Berman & Sumida 1990; Sumida & Berman 1993) has been confidently assigned a Late Pennsylvanian age (Vaughn 1969, 1972). El Cobre Canyon in north-central New Mexico preserves a significant section of the Permo-Pennsylvanian sequence (Eberth & Miall 1991; Berman 1993). It is included here as a Late Pennsylvanian representative, as the localities considered here occur in the Pennsylvanian section of the Canyon. El Cobre Canyon is actually a series of closely situated localities which may well represent more than a single palaeoenvironment, but its inclusion here is rationalised due to the proximity of the localities in the canyon. Arroyo de Agua in north-central New Mexico has been similarly interpreted as spanning the Permo-Pennsylvanian boundary, although those taxa enumerated here have been assigned to a position near the Leonardian–Wolfcampian boundary (Berman *et al.* 1987, 1988). Although the age of the Halgaito Shale has been debated, the overlying Organ Rock Sandstone is confidently interpreted as Early Permian in age (Vaughn 1964, 1973; Sumida *et al.* 1998a, c, d). As most of the localities found in the Organ Rock Sandstone of southeastern Utah and northern Arizona have been interpreted as having similar lithologies and environments of deposition (Stanesco & Campbell 1989; Sumida *et al.* 1998a, c, d), they are also considered together. The Lower Permian placement of the Placerville, Colorado locality is based on the assignment

by Lewis & Vaughn (1965). Although a small assemblage, it is included here because of its rather restricted spatial distribution. The Lower Permian Rattlesnake Canyon, Archer City, and the Geraldine Bonebed localities are recognised here as a series of geographically restricted localities from north-central Texas. Sander's (1987, 1989) taphonomic analyses of these localities provide both taxonomic lists and palaeoenvironmental interpretations of these sites (also see Hook 1989). Olson (1977) compared the Lower Permian lake and more terrestrial localities of Waurika and Orlando, respectively, in central Oklahoma. As these localities provided the impetus for one of the earliest models of Late Palaeozoic food webs, they are important elements in any locality-by-locality comparisons. The fissure fills in the Ordovician Arbuckle Limestone at the Dolese Brothers Quarry, Richards Spur (commonly referred to as Fort Sill), Oklahoma have generally been accepted as Leonardian in age (Sullivan & Reisz 1999). Fort Sill records the greatest taxonomic diversity known for a vertebrate assemblage from the Lower Permian (Sullivan & Reisz 1998). A small assemblage is known from Prince Edward Island, Canada (Langston 1963). Although Langston (1963) assigned the red-beds studied there only tentatively to the Lower Permian, they are included here in an effort to clarify their stratigraphic position. Localities from the tri-state area of Ohio, West Virginia, and Pennsylvania are conspicuously absent in this survey. Although of potentially comparable age, specimens recovered in this region are typically fragmentary, and from widely scattered localities representing a diversity of environments.

Interpretations of organisms as terrestrial, semi-aquatic, or aquatic are frequently made on the basis of anatomical structure. Sumida (1997) provided examples of such an analysis, where locomotor features and the degree of limb bone ossification provide criteria for the recognition of degrees of terrestriality. However, such methods are not adequate when attempting comparative palaeoenvironmental analyses. Using anatomical criteria as a basis of palaeoenvironmental reconstruction, and then declaring associated taxa as terrestrially adapted verges on circular reasoning. Thus, when possible, palaeoenvironmental interpretations for individual localities should be based on sedimentological data independent of the morphology of the fossils themselves. The Bromacker locality satisfies this condition (Eberth *et al.* 2000). Not all of the localities considered here have been independently analysed in a strictly sedimentological context, but enough of them have to allow the development of an independent palaeoenvironmental context for Late Palaeozoic localities. Table 2 summarises those localities for which such sedimentological data are available. The palaeoenvironments surveyed range from predominantly pond and lake deposits, to strictly terrestrial deposits.

Assemblage comparisons

Table 3 provides a locality-by-locality comparison of the Bromacker with Late Palaeozoic assemblages in the USA and Canada. With the exception of the Bromacker, all of the assemblages include some aquatic or semi-aquatic taxa.

Late Pennsylvanian assemblages

Among the Late Pennsylvanian assemblages, the Halgaito Shale of southeastern Utah preserves the greatest number of fish taxa, although El Cobre Canyon and Badger Creek contain lungfish and palaeoniscoids, respectively. Badger Creek and the Halgaito also show evidence of aquatic lepospondyls. Although ubiquitous in Early Permian assemblages, *Seymouria* is not common in those of the Late Pennsylvanian. Only one very tentatively identified seymouriid is known from the Halgaito, based only on a partial vertebra (Sumida *et al.* 1998d). The Bromacker is well known for its excellent preservation of a number of specimens of *Diadectes*, as well as a new relative of that genus, *Orobates* Berman *et al.*, 2003. Although all Pennsylvanian localities considered here include diadectomorphs, they are either the more basal *Limnoscelis* or diadectids other than *Diadectes*. A tentative hypothesis of relationships of diadectomorphs by Kissel & Reisz (2002) indicates that the Pennsylvanian genera are more basal forms than *Diadectes*. Pelycosaurian-grade synapsids are found at all the Late Pennsylvanian localities surveyed, but edaphosaurids and semi-aquatic ophiacodonts are much more frequently encountered. With the exception of the tenuous presence of *Seymouria* in southeastern Utah, the Bromacker shares virtually no taxa in common with any of the Late Pennsylvanian localities surveyed.

Early Permian assemblages

Not unexpectedly, the Bromacker vertebrate assemblage shares more in common with those identified as Early Permian in age than those of Late Pennsylvanian age. However, in all cases the Bromacker assemblage compares only in part with each of the Early Permian assemblages surveyed.

Xenacanth sharks are found at most of the Early Permian localities, as are actinopterygian fish, mainly palaeoniscoids. Of Late Palaeozoic lungfish, *Gnathoriza* is commonly considered to have been capable of surviving periods of severe drought by aestivating in burrows, whereas *Sagenodus* was not. The presence of the latter at Archer City, Geraldine Bonebed, and Waurika is consistent with their environmental interpretations as oxbow lakes or swamps that dried completely less frequently (Sander 1987, 1989; Olson 1977). Of the localities surveyed, only the assemblages at Prince Edward Island, Fort Sill, and Bromacker, lack fish completely.

Of amphibians[1], *Eryops* is found at every Early Permian locality surveyed except for the Bromacker. *Zatrachyes* is also somewhat common, as it is absent from only southeastern Utah, Placerville, Prince Edward Island, Fort Sill, and Bromacker. Both *Eryops* and *Zatrachyes* are generally considered to be semi-aquatic amphibians (Olson 1977; however see Pawley 2002 for a dissenting view regarding *Eryops*), and are frequently found well preserved in pond deposits (Sander 1987, 1989; Frede *et al.* 1993). Among dissorophoid amphibians, trematopids and dissorophids are considered to be highly adapted to a terrestrial existence (Olson 1970; Sumida 1997). Although not uncommon in the Late Palaeozoic in general, the terrestrial trematopids are not common at the localities surveyed here. Two distinct trematopids are found at Fort Sill, and one each at Archer City and Bromacker. Dissorophids are found at Arroyo del Agua (two), Placerville (one), Fort Sill (two), and Bromacker (one). Among the most widely distributed of amphibians is the seymouriamorph *Seymouria*. *Seymouria* is a terrestrially adapted tetrapod (Sumida 1997) found at all but three of the Lower Permian sites. Whereas limnoscelids and diadectids are the predominant diadectomorphs at Late Pennsylvanian localities, limnoscelids are not found at Early Permian localities and the diadectids found are predominantly *Diadectes*, a genus seemingly restricted to the Early Permian. The identification of limnoscelids at Arroyo de Agua (Langston 1966) and Placerville (Lewis & Vaughn 1965), which would suggest older age assignments for these two localities, has recently been refuted by Wideman (2002). In the wake of limnoscelid absence, tseajaiids are present at some of the Early Permian localities. However, *Diadectes* becomes the single most common diadectomorph, being found at all of the Early Permian localities surveyed. It is noteworthy that Fort Sill and Bromacker have both yielded a second diadectid, *Orobates* at Bromacker and an as yet unnamed one at Fort Sill, and they may be very similar to one another (Reisz & Sutherland 2001).

While present but in low numbers of individual specimens, the pelycosaurian-grade synapsid ophiacodonts and edaphosaurs are common to ubiquitous at the Early Permian localities surveyed. The only exceptions are Fort Sill and Bromacker. No representatives of these groups – which are generally interpreted as semi-aquatic or closely tied to aquatic environments – are found at Fort Sill or Bromacker. Alternatively, both Fort Sill and Bromacker preserve caseid and varanopid pelycosaurs. Caseids are commonly interpreted as terrestrial, high-fibre herbivores (Sues & Reisz 1998), whereas varanopids are probably terrestrial predators (Reisz *et al.* 1998). No sphenacodontids are known from Fort Sill, but Bromacker provides the only record of the sphenacodontid *Dimetrodon* in Europe.

Table III. Comparison of vertebrate assemblages from selected Upper Pennsylvanian and Lower Permian localities.

	Upper Carboniferous/Pennsylvanian			Lower Permian									
	Halgaito Shale Utah	El Cobre Canyon New Mexico	Badger Creek Colorado	Organ Rock SS Utah	Arroyo del Agua New Mexico	Placerville Colorado	Rattlesnake Canyon Texas	Archer City Texas	Geraldine Bonebed Texas	Waurika Oklahoma	Prince Edward Island	Fort Sill Oklahoma	Bromacker Germany
Chondrichthyes	*Orthacanthus*			Xenacanth	*Xenacanthus*		*Orthacanthus*	*Orthacanthus*	*Orthacanthus*	*Xenacanthus*			
Actinopterygii	Palaeoniscoid	Palaeoniscoids		Palaeoniscoid Phyllodont	*Progyrolepis*		Palaeoniscoids	Palaeoniscoids		Ganoid			
Dipnoi	*Gnathoriza* *Sagenodus*	*Gnathoriza*						*Sagenodus*	*Sagenodus*	*Sagenodus*			
Crossopterygii							*Spermatodus*	*Ectosteorhachis*	*Ectosteorhachis*				
Nectridia	*Lohsania* *Diplocaulus*									*Diplocaulus*			
Aïstopoda	*Phlegethontia*											*Phlegethontia*	
Microsauria/Tuditanomorpha			*Coloraderpeton* *Trihecaton*		*Stegotretus*					*Pantylus*			
Temnospondyli		*Chenoprosopus*	?		*Chenoprosopus*								
Trimerorhachoidea	Trimerorhachid		Trimerorhachid	Trimerorhachid			*Trimerorhachis*		*Trimerorhachis*	*Trimerorhachis*	*Trimerorhachis*		
Edopoidea													
Eryopoidea	*Eryops*	*Eryops*	*Eryops*	*Eryops*	*Eryops*	*Eryops*	*Eryops*	*Eryops*	*Eryops*	*Eryops*	*Eryops*	Eryopid	
Zatracheidae			Zatracheid	Zatracheid	*Zatrachys*		*Zatrychys*	*Zatrychys*	*Zatrychys*	*Zatrychys*			
Dissorphoidea	*Platyhystrix*	*Aspidosaurus* *Platyhystrix*	*Platyhystrix*		*Broiliellus* *Ecolosonia* *Platyhystrix*	*Platyhystrix*						*Cacops* *Tersomius*	New taxon
Trematopidae		*Anconastes*						*Acheloma*				*Acheloma* Trematopid	*Tambachia*
Anthracosauria Embolomeri	*Archeria*	Embolomere					*Archeria*	*Archeria*	*Archeria*	*Archeria*			
Seymouriamorpha	*Seymouria?*	*Seymouria*		*Seymouria*	*Seymouria*	Seymouriid		*Seymouria*			*Seymouria*	*Seymouria*	*Seymouria*
Diadectomorpha Limnoscelidae	Limnoscelid	*Limnoscelis*	*Limnoscelis*										
Tseajaiidae				*Tseajaia*	*Tseajaia*								
Diadectidae	Diadectid	*Diasparactus*	*Desmatodon*	*Diadectes*	*Diadectes*	*Diadectes*	*Diadectes*	*Diadectes*	*Diadectes*	*Diadectes*	Diadectid	*Diadectes* New taxon	*Diadectes* New taxon
Synapsida Caseosauria		*Aerosaurus* *Nitosaurus*			*Aerosaurus* *Oedaleops*							Caseid	Caseid
Eupelycosauria Varanopsidae						*Mycterosaurus*					*Mycterosaurus*	*Mycterosaurus*	Varanopid
Ophiacodontidae	*Ophiacodon*	*Ruthromia* *Ophiacodon* *Baldwinonus*	Ophiacodontid	*Ophiacodon*	*Ophiacodon*	*Ophiacodon*		*Ophiacodon*	*Ophiacodon*	*Ophiacodon*	*Ophiacodon*		
Edaphosauridae	*Edaphosaurus*	*Edaphosaurus*	*Ianthasaurus* New taxon		*Edaphosaurus*		*Edaphosaurus*	*Edaphosaurus*	*Edaphosaurus*	*Edaphosaurus*	*Trichasaurus*		
Sphenacodontidae	*Sphenacodon*	*Sphenacodon*	Haptodontid Sphenacodontid	*Ctenospondylus* *Dimetrodon* *Sphenacodon*	*Sphenacodon*	*Cutleria*	*Dimetrodon*	*Dimetrodon*	*Dimetrodon*	*Sphenacodon*	*Bathygnathus*		*Dimetrodon*
Reptilia Pararaptilia/Bolosauridae					*Bolosaurus?*		*Bolosaurus*	*Bolosaurus*	*Bolosaurus*	*Bolosaurus?*		*Bolosaurus*	*Eudibamus*
Captorhinidae		*Chamasaurus?*			*Rhiodenticulatus* Captorhinid	Captorhinid		*Romeria*		?		Captorhinid	*Thuringothyris*
Protorothyrididae			Protorothyridid									*Colobomycter*	
Araeoscelidia/Diapsida	Araeoscelid				*Zarcasaurus*							Neodiapsid	

Discussion and analysis

Biostratigraphy

Seymouria, *Diadectes*, and possibly *Dimetrodon*, stand out as the most biostratigraphically useful of the taxa common in Early Permian assemblages. With the exception of one questionable identification from the Halgaito of southeastern Utah, *Seymouria* is restricted to the Early Permian, so this age can be confidently assigned to the Bromacker assemblage. It is notable that the species found at Bromacker, *Seymouria sanjuanensis*, is the same as that found in southeastern Utah (Vaughn 1966) and northern New Mexico (Berman *et al.* 1987), providing even greater resolution. This has led Sumida *et al.* (1996) and Berman *et al.* (2000) to suggest that the Bromacker assemblage is earliest Permian (Wolfcampian) in age. The absence of *Seymouria* from Leonardian localities in Texas (Hook 1989), as well as from some upper Wolfcampian localities such as the Geraldine Bonebed and even Archer City, supports Lucas' (2002) contention that the genus provides a useful biostratigraphic marker for non-marine localities of the earliest Permian. The combined global distribution and narrow temporal restriction of *Seymouria sanjuanensis* to the earliest Permian Wolfcampian further suggests a strong biostratigraphic utility of this species, particularly if found at other localities in the future.

Diadectomorphs are found at all of the Late Pennsylvanian and Early Permian localities surveyed. Late Pennsylvanian localities preserve either *Limnoscelis* or diadectids exclusive of *Diadectes* that include the more basal forms *Desmatodon* or *Diasparactus*. *Diadectes* is found at every Lower Permian locality surveyed here. Thus, whereas it may afix an Early Permian age to a locality, its range through the Lower Permian may render it somewhat less useful than *Seymouria* as a terrestrial index fossil of Permian time scales (Lucas 2002). *Eudibamus* was based on a very well-preserved post-cranium but poorly preserved skull. Given that most records of the genus *Bolosaurus* are based on cranial materials but a paucity of post-cranial materials, the Bolosauridae is less reliable for drawing biostratigraphic and palaeoenvironmental conclusions. However, if more complete materials combining cranial, dental, and post-cranial components are found, this limitation could change.

Palaeoenvironments

Although the Bromacker locality has been interpreted as a strictly terrestrial Early Permian ecosystem (Eberth *et al.* 2000), much of its comparative analysis had been made on a regional basis. Here, a more rigorous comparison of localities and palaeoenvironmental hypotheses reconfirms that model.

- No fishes of any kind have been found at Bromacker in over a quarter of a century of collecting (Berman *et al.* 2001).

- Those taxa regarded as anamniote "amphibians" are, nonetheless, terrestrially adapted groups. The seymouriamorph *Seymouria* does not demonstrate aquatic larval forms even in the smallest specimens recovered at Bromacker. Notably, the smallest of these specimens (Berman & Martens 1993) is smaller than the largest larval specimen of the closely related discosauriscids found in more typical limnetic localities in central Germany. This demonstrates the morphological and palaeoenvironmental distinction of these taxa. Dissorophid and trematopid amphibians also exhibit terrestrial adaptations (Sumida *et al.* 1998d).

- Bromacker clearly establishes high-fibre herbivores as a viable base for a terrestrial food web (Eberth *et al.* 2000). Such high-fibre herbivory is characteristic of terrestrially adapted amniotes or their near relatives (Hotton *et al.* 1997; Sues & Reisz 1998; Berman *et al.* 1998). High-fibre herbivores at Bromacker include numerous specimens of the diadectomorph *Diadectes* and *Orobates*, a primitive caseid of pelycosaurian-grade synapsids, and the small bolosaurid *Eudibamus*.

- Of pelycosaurian-grade synapsids found at Bromacker, the sphenacodontid *Dimetrodon teutonis* and a potentially new varanopid are known in addition to the caseid. Sphenacodontids and varanopids are terrestrially adapted in terms of their locomotor features (Sumida 1997) and their preferred prey (Romer & Price 1940; Reisz 1986).

- Given its interpretation as a cursorial biped, there is little doubt that the bolosaurid reptile *Eudibamus* was not only adapted in terms of terrestrial feeding, but in terms of its locomotor adaptations as well (Berman *et al.* 2000).

- All of the taxa listed above are extremely well preserved, exhibiting in most cases little or no transport, indicating that death and burial were probably coeval events. This, combined with evidence of completely draped palaeochannel dunes (Eberth *et al.* 2000), suggests that, although flowing water may have been intermittently present, it was probably directed to an internally drained basin, in contrast to the fluvial systems more typical of North American localities.

- Subaerial exposure of standing water was probably of short duration (Eberth *et al.* 2000), thus preventing colonisation of the Bromacker by aquatic vertebrates. Palaeontological, sedimentological, and taphonomic analyses of the Bromacker clearly establish its vertebrate assemblage as strictly terrestrial in nature.

Other than the Bromacker, Fort Sill, Oklahoma stands out as the only other terrestrially dominated assemblage locality of those surveyed here. Only two taxa traditionally considered to be potentially aquatic in nature are present at Fort Sill, the aistopod *Phlegethontia* and the temnospondyl *Eryops*. The record of *Eryops* at Fort Sill is based on a single partial dermal element (Olson 1991), and may represent an "erratic" component of the assemblage. *Phlegethontia* has traditionally been considered to be an aquatic amphibian (McGinnis 1967), but Anderson's (2002) recent restudy of the Phlegethontiidae reveals no specifically aquatic adaptations in the genus. With the domination of the Fort Sill assemblage by the terrestrial captorhinid reptile *Captorhinus*, and only the possibly erratic *Eryops* as an aquatic form, Fort Sill stands as the most diverse of terrestrially dominated faunal assemblages. Although the Fort Sill assemblage is larger than that of Bromacker, the palaeoenvironmental interpretation of the latter is supported by rigorous sedimentological analysis as well. Only a preliminary hypothesis of the environment of deposition at Fort Sill has been forwarded at this point. Reisz & Sutherland (2001) suggested that Fort Sill is the earliest record of a palaeokarst. Although this hypothesis remains to be more thoroughly tested, it is important to note that, if correct, it reinforces the interpretation of Fort Sill as a terrestrial locality as karsts develop only in subareal limestones.

The terrestrially adapted Bromacker assemblage allows the "dissection" of strictly terrestrial components out of other mixed assemblages, as well as a qualitative estimation of the degree of terrestriality exhibited by other localities (Sumida *et al.* 2003). The examination of other Late Palaeozoic localities suggests that a number of taxa not strictly defined as aquatic may nonetheless be closely tied to aquatic environments. Olson (1977) noted that the lake-margin faunas may have had characteristic elements, although they would be difficult to identify with certainty in mixed assemblages. The removal of strictly terrestrial forms as exemplified by the Bromacker assemblage representatives on the one hand, and clearly aquatic forms such as fishes and aquatic amphibians on the other hand, provides a tool for the identification of such intermediate, semi-aquatic forms. The most likely candidates for such ecologically intermediate forms would be the temnospondyl amphibian *Eryops* and its close relatives, and ophiacodont and edaphosaurid pelycosaurs.

Acknowledgements

The authors thank Dr Gavin Young for the invitation to contribute this study to the proceedings of the Early Vertebrate Symposium held at the International Palaeontological Congress in Sydney, Australia in July 2002. This study was supported by grants from the National Geographic Society, North Atlantic Treaty Organization (NATO), the Edward O'Neil Endowment Fund, the M. Graham Netting Fund, Carnegie Museum of Natural History, and California State University San Bernardino College of Natural Sciences. The authors thank Drs Jason Anderson, Sean Modesto, Elizabeth Rega, and Robert Reisz for helpful discussion and suggestions, and Drs Anne Warren and Gavin Young for helpful comments on the manuscript. Dr Elizabeth Rega provided critical translation services while in Germany and aided in measuring stratigraphic sections. Ms Heike Sheffel provided logistical support during fieldwork in Germany.

Note

1. Although understood as a paraphyletic term, "amphibian" is used here in a colloquial sense as equivalent to anamniote tetrapod.

References

Anderson, J.S. 2002: Revision of the aïstopod genus *Phlegethontia* (Tetrapoda: Lepospondyli). *Journal of Paleontology* 76, 1029–1046.

Baars, D.L. 1962: Permian system of the Colorado Plateau. *American Association of Petroleum Geologists Bulletin* 46, 149–218.

Baars, D.L. 1991: Redefinition of the Pennsylvanian–Permian boundary in Kansas, Midcontinent, U.S.A. Program with Abstracts, International Congress of the Permian System of the World, Perm USSR, p. A3.

Baars, D.L. 1995: *Navajo Country, a Geology and Natural History of the Four Corners Region.* University of New Mexico Press, Albuquerque.

Baker, A.A. 1936: Geology of the Monument Valley–Navajo Mountain region, San Juan County, Utah. *United States Geological Survey Bulletin* No. 865, 106 pp.

Berman, D.S. 1993: Lower Permian vertebrate localities of New Mexico and their assemblages. *In* Lucas, S.G. & Zidek, J. (eds): *Vertebrate Paleontology in New Mexico*, 11–21. *New Mexico Museum of Natural History and Science Bulletin*, No. 2.

Berman, D.S., Eberth, D.A. & Brinkman, D.B. 1988: *Stegotretus agyrus*, a new genus and species of microsaur (Amphibia) from the Permo-Pennsylvanian of New Mexico. *Annals of Carnegie Museum 57*, 293–323.

Berman, D.S., Henrici, A.C., Kissel, R.A., Sumida, S.S. & Martens, T. 2003. A new diadectid (Diadectomorpha), *Orobates pabsti*, from the Early Permian of central Germany. *Bulletin of the Carnegie Museum of Natural History* (in press).

Berman, D.S., Henrici, A.C., Sumida, S.S. & Martens, T. 2000a: Redescription of *Seymouria sanjuanensis* (Seymouriamorpha) from the Lower Permian of Germany based on complete, mature specimens with discussion of the paleoecology of Bromacker locality assemblage. *Journal of Vertebrate Paleontology* 20, 253–268.

Berman, D.S. & Martens, T. 1993: First occurrence of *Seymouria* (Amphibia: Batrachosauria) in the Lower Permian of central Germany. *Annals of Carnegie Museum* 62, 63–79.

Berman, D.S., Reisz, R.R. & Eberth, D.A. 1987: *Seymouria sanjuanensis* (Amphibia, Batrachosauria) from the Lower Permian Cutler Formation of north-central New Mexico and the occurrence of sexual dimorphism in that genus questioned. *Canadian Journal of Earth Sciences* 24, 1769–1784.

Berman, D.S., Reisz, R.R., Martens, T. & Henrici, A.C. 2001: A new species of *Dimetrodon* (Synapsida: Sphenacodontidae) from the Lower Permian of Germany records first occurrence of genus outside of North America. *Canadian Journal of Earth Sciences* 38, 803–812.

Berman, D.S., Reisz, R.R., Scott, D., Henrici, A.C., Sumida, S.S. & Martens, T. 2000b: Early Permian bipedal reptile. *Science* 290, 969–972.

Berman, D.S. & Sumida, S.S. 1990: A new species of *Limnoscelis* (Amphibia, Diadectomorpha) from the Late Pennsylvanian Sangre de Cristo Formation of central Colorado. *Annals of Carnegie Museum* 59, 303–341.

Berman, D.S, Sumida, S.S. & Martens, T. 1998: *Diadectes* (Diadectomorpha: Diadectidae) from the Early Permian of central Germany, with description of a new species. *Annals of Carnegie Museum* 67, 53–93.

Boy, J.A. & Martens, T. 1991: A new captorhinomorph reptile from the Rotliegend of Thuringia (Lower Permian; eastern Germany). *Paläontologische Zeitschrift* 65, 363–389.

Chernykh, V.V., Ritter, S.M. & Wardlaw, B.R. 1997: *Streptognathus isolatus* new species (Conodonta): proposed index for the Carboniferous–Permian boundary. *Journal of Paleontology* 71, 162–164.

Davydov, V.I., Glenister, B.F., Spinosa, C., Ritter, S.M., Chernykh, V.V., Wardlaw, B.R. & Snyder, W.S. 1995: Proposal of Aidaralash as GSSP for the base of the Permian system. *Permofiles* 26, 1–9.

Eberth, D.A. & Berman, D.S. 1993: Stratigraphy, sedimentology, and vertebrate paleoecology of the Cutler Formation redbeds (Pennsylvanian–Permian) of north-central New Mexico. *In* Lucas, S.G. & Zidek, J. (eds): *Vertebrate Paleontology in New Mexico*, 33–48. New Mexico Museum of Natural History and Science Bulletin, No. 2.

Eberth, D.A., Berman, D.S., Sumida, S.S. & Hopf, H. 2000: Lower Permian terrestrial paleoenvironments and vertebrate paleoecology of the Tambach Basin (Thuringia, central Germany): the upland Holy Grail. *Palaios* 15, 293–313.

Eberth, D.A. & Miall, A.D. 1991: Stratigraphy, sedimentology, and evolution of a vertebrate-bearing, braided to anastomosing fluvial system, Cutler Formation (Pennsylvanian–Permian), north-central New Mexico. *Sedimentary Geology* 72, 225–252.

Frede, S., Sumida, S.S. & Berman, D.S. 1993: New information on Early Permian vertebrates from the Halgaito tongue of the Cutler Formation of southeastern Utah. *Journal of Vertebrate Paleontology* 13, 36A.

Hook, R.W. 1989: Stratigraphic distribution of tetrapods in Bowie and Wichita Groups, Permo-Carboniferous of north-central Texas. *In* Hook, R.W. (ed.): *Permo-Carboniferous Paleontology, Lithostratigraphy, and Depositional Environments of North-Central Texas. Field Trip Guidebook 2*, 47–53. Annual Meeting of the Society of Vertebrate Paleontology, Austin, Texas.

Hotton, N., Olson, E.C. & Beerbower, R. 1997: The amniote transition and the discovery of herbivory. *In* Sumida, S.S. & Martin, K.L.M. (eds): *Amniote Origins, Completing the Transition to Land*, 207–264. Academic Press, San Diego.

Kissel, R.A., Berman, D.S., Henrici, A.C., Reisz, R.R., Sumida, S.S. & Martens, T. 2002: A new diadectid (Tetrapoda: Diadectomorpha) from the Lower Permian of Germany. *Journal of Vertebrate Paleontology* 22, 74A.

Kissel, R.A. & Reisz, R.R. 2002: Remains of a small diadectid (Tetrapoda: Diadectomorpha) from the Dunkard Group of Ohio, with consideration of diadectomorph phylogeny. *Journal of Vertebrate Paleontology* 23, 67A.

Langston, W. 1963: Fossil vertebrates and the Late Palaeozoic red beds of Prince Edward Island. *Bulletin of the National Museum of Canada*, No. 187, 36 pp.

Langston, W. 1966: *Limnosceloides brachycoles* (Reptilia: Captorhinomorpha), a new species from the Lower Permian of New Mexico. *Journal of Paleontology* 40, 690–695.

Lewis, G.E. & Vaughn, P.P. 1965: Early Permian vertebrates from the Placerville area, Colorado. *Contributions to Paleontology, Geological Society Professional Paper 503-C*, 46 pp.

Lucas, S.G. 2002: Tetrapods and the subdivision of Permian time. *In* Hills, L.V., Henderson C.M. & Bramber, E.W. (eds): *Carboniferous and Permian of the World*, 479–491. Canadian Society of Petroleum Geologists Memoir 19.

Martens, T. 1988: Die Bedeutung der Rotsedimente für die Analyse der Lebewelt des Rotliegenden. *Zeitschrift für Geologische Wissenshaften* 16, 933–938.

McGinnis, H.J. 1967: The osteology of *Phlegethontia*, a Carboniferous and Permian aïstopod amphibian. *University of California Publications in Geology* 71, 1–46.

Olson, E.C. 1970: *Trematops stonei* sp. nov. (Temnospondyli: Amphibia) from the Washington Formation, Dunkard Group, Ohio. *Kirtlandia* 8, 1–12.

Olson, E.C. 1977: Permian lake faunas: a study in community evolution. *Journal of the Palaeontological Society of India* 20, 146–163.

Olson, E.C. 1991: An eryopid (Amphibia; Labyrinthodontia) from the Fort Sill fissures; Lower Permian, Oklahoma. *Journal of Vertebrate Paleontology* 11, 130–132.

Pabst, W. 1896: Thierfahrten aus dem Oberrrotliegenden von Tambach in Thüringen. *Zeitschrift der Deutschen Geologischen Gesellschaft* 47, 570–576.

Pawley, K. 2002: The postcranial skeleton of temnospondyl amphibians. *Journal of Vertebrate Paleontology* 22, 95A–96A.

Reisz, R.R. 1986: Pelycosauria. *Handbuch der Paläeoherpetologie*, Tiel 17A. Gustav Fischer, Stuttgart.

Reisz, R.R., Dilkes, D.W. & Berman, D.S. 1998: Anatomy and relationships of *Elliotsmithia longiceps* Broom, a small synapsid (Eupelycosauria, Varanopseidae) from the Late Permian of South Africa. *Journal of Vertebrate Paleontology* 18, 602–611.

Reisz, R.R. & Sutherland, T.E. 2001: A diadectid (Tetrapoda: Diadectomorpha) from the Lower Permian fissure fills of the Dolese Quarry, near Richards Spur, Oklahoma. *Annals of Carnegie Museum* 70, 133–142.

Romer, A.S. & Price, L.I. 1940: Review of the Pelycosauria. *Geological Society of America Special Paper 28*, 1–538.

Sander, M. 1987: Taphonomy of the Lower Permian Geraldine bonebed in Archer County, Texas. *Palaeogeography, Palaeoclimatology, Palaeoecology* 61, 221–236.

Sander, M. 1989: Early Permian depositional environments and pond bonebeds in central Archer County, Texas. *Palaeogeography, Palaeoclimatology, Palaeoecology* 69, 1–21.

Stanesco, J.D. & Campbell, J.A. 1989: Eolian and noneolian facies of the Lower Permian Cedar Mesa Sandstone Member of the Cutler Formation, southeastern Utah. *United States Geological Survey Bulletin 1808*, F1–F13.

Sues, H.-D. & Reisz, R.R. 1998: Origins and early evolution of herbivory in tetrapods. *Trends in Ecology and Evolution* 13, 141–145.

Sullivan, C. & Reisz, R.R. 1999: First record of *Seymouria* (Vertebrata: Seymouriamorpha) from Early Permian fissure fills at Richards Spur, Oklahoma. *Canadian Journal of Earth Science* 36, 1257–1266.

Sumida, S.S. 1997: Locomotor features of taxa spanning the amphibian to amniote transition. *In* Sumida, S.S. & Martin, K.L.M. (eds): *Amniote Origins, Completing the Transition to Land*, 353–398. Academic Press, San Diego.

Sumida, S.S., Albright, G.A. & Rega, E.A. 1998a: Late Paleozoic fishes of Utah. *In* Gillette, D.D. (ed.): *Vertebrate Paleontology in Utah*, 13–20. Miscellaneous Publication 99-1, Utah Geological Survey.

Sumida, S.S. & Berman, D.S. 1993: The pelycosaurian (Amniota: Synapsida) assemblage from the Late Pennsylvanian Sangre de Cristo Formation of central Colorado. *Annals of Carnegie Museum* 62, 293–310.

Berman, D.S., Henrici, A.C., Kissel, R.A., Eberth, D.A. & , 2002: Origin of the modern terrestrial vertebrate ecosys-nented by an Early Permian assemblage from Germany. *ertebrate Paleontology 22*, 112A–113A.

Berman, D.S., Henrici, A.C. & Martens, T. 2001: ian-grade synapsids from the Lower Permian of central the apex of an exclusively terrestrial foodweb. *PaleoBios* 122–123.

Berman, D.S. & Martens, T. 1996: Biostratigraphic s between the Lower Permian of North America l Europe using the first record of an assemblage of etrapods from Germany. *PaleoBios 17*, 1–12.

Berman, D.S. & Martens, T. 1998b: A new trematopid am-m the Lower Permian of central Germany. *Palaeontology* 9.

Eberth, D.A. & Berman, D.S. 1999: 'Dissecting' a Late ertebrate paleoenvironment: identification of a truly ter-system component. *Journal of Vertebrate Paleontology 19*,

Eberth, D.A., & Berman, D.S. 2000: Refining the concept e Paleozoic chronofauna: Early Permian vertebrates of t exclusively terrestrial ecosystem. *Journal of Vertebrate* gy 20, 72A.

Lombard, R.E., Berman, D.S. & Henrici, A.C. 1998c: Late umniotes and the near relatives from Utah and northeast-a, with comments on the Permian–Pennsylvanian bound-n and northern Arizona. *In* Gillette, D.D. (ed.): *Vertebrate*

Paleontology in Utah, 31–43. Miscellaneous Publication 99-1, Utah Geological Survey.

Sumida, S.S., Walliser, J.B.D. & Lombard, R.E. 1998d: Late Paleozoic amphibian-grade tetrapods of Utah. *In* Gillette, D.D. (ed.): *Vertebrate Paleontology in Utah*, 21–30. Miscellaneous Publication 99-1, Utah Geological Survey.

Vaughn, P.P. 1962: Vertebrates from the Halgaito Tongue of the Cutler Formation, Permian of San Juan County, Utah. *Journal of Paleontology 36*, 529–539.

Vaughn, P.P. 1964: Vertebrates from the Organ Rock Shale of the Cutler Group, Permian of Monument Valley and vicinity, Utah and Arizona. *Journal of Paleontology 38*, 567–583.

Vaughn, P.P. 1966: *Seymouria* from the Lower Permian of southeastern Utah, and possible sexual dimorphism in that genus. *Journal of Paleontology 40*, 603–612.

Vaughn, P.P. 1969: Upper Pennsylvanian vertebrates from the Sangre de Cristo Formation of central Colorado. *Contributions in Science, Los Angeles County Museum of Natural History 164*, 1–28.

Vaughn, P.P. 1972: More vertebrates, including a new microsaur, from the Upper Pennsylvanian of central Colorado. *Contributions in Science, Los Angeles County Museum of Natural History 223*, 1–30.

Vaughn, P.P. 1973: Vertebrates from the Cutler Group of Monument Valley and vicinity, Utah and Arizona. *Guidebook of Monument Valley and Vicinity: New Mexico Geological Society Field Conference Volume 24*, 99–105.

Wideman, N. 2002: The postcranial anatomy of the Late Paleozoic family Limnoscelidae and its significance for diadectomorph taxonomy. *Journal of Vertebrate Paleontology 22*, 119A.